D0519148

27 MAR 1999

1 5 APR 1999

− 7 MAY 1999

3 MAR 2006

− 7 NOV 2008

−5 DEC 2008

# FRED DIBNAH STEAMSON

by
Fred Dibnah
with Peter Nicholson

Line
one
publishing ltd

485005173

ᒪOOO3O6568

Published by
Line One Publishing Limited
Bayshill House
Bayshill Road
Cheltenham
Glos.
GL50 3AA


ISBN 0 907036 20 1

All photographs are by Peter Nicholson
unless stated otherwise.
Grateful thanks are due to the following for
the supply of photographs: Queen St. Mill;
British Telecom (Gloucester Telephone Area)
and Daily Mail (Associated Newspapers).

# Contents

*Fred and Alison Dibnah's now famous 12 ton steam roller, with Fred at the controls. Named "Alison", it is Aveling & Porter No.7632, built in February 1912 and carries registration number DM 3079.*

## Chapter 1

# Alison—The Steam Roller

In 1966 I bought a steam roller that were absolutely knackered. The rear wheels sloped inwards and the water tank at the back was full of holes, all the bearings were shot out of it and you could see daylight through the funnel and the smokebox.

This was Aveling & Porter No. 7632 built in February 1912, and represented something I had longed to own for many years, and so despite its condition, had jumped at the chance to buy it. I bought this specimen off two Welsh scrap iron merchants for £175, not that I ever got a receipt for it. They dragged it out of a shed and onto a country lane and just left it there, beside the road. All 12 tons of it, most of which appeared to be rust. I gave them the money and off they went leaving me to it.

Fortunately for me I knew this farmer who had an 1899 Foden traction engine and he agreed to tow it from the old army camp where we had found it, to his farm which was a distance of about seven miles. We eventually got there, not having had too much experience with steam engines. I had steered one or two of them before and that were about it really. We then played about with it for a couple of weeks until this chap, a sort of an engineer who used to come and fettle up the farmer's engine from time to time, came along and gave it a look over.

He told us what he thought were wrong with it, he told us to change this and do that before we should attempt the journey home to Bolton, which were a distance of about 9 miles. We did what he said and then on the next Sunday morning we arrived at the farm at about 7.00 am and lit the fire. The engineer chap agreed to drive the engine, it being the first time it had travelled under its own steam for nearly 20 years. This was a big moment as we proudly set off, bound for Bolton.

We got about 4 miles down the road and the thing conked out on us. However, in a matter of three or four weeks we had learnt quite a bit about steam rollers and realised that the trouble was with the injector, the thing that puts the water into the boiler. This were completely blocked up with muck so we removed it and took it home and had a good look at it. We then took it round to the local brass foundry for some expert attention. The fellow there had a quick look at it and said,

"I don't want to take your money off you cock. So take it home, throw it in the fire, then take it outside and bang it on the flags. These brass cones will fall out and all the muck'll fall out".

We did all that and returned to the engine the following weekend, refitted the injector and lit the fire. Beautiful!

This was before I was married, though Alison came with me and together with a friend, Andrew Shorrocks, we managed to get the engine back to Bolton. We arrived one Sunday afternoon and left the roller on the main street, in front of a row of terraced houses. It was nearly all rust, but you could see a few traces of green paint here and there. We then had to find somewhere to keep it and to work on it.

A friend of mine had a hen pen and he had promised me that if I ever got an engine he would let me leave it there. However, he had gone and died in the meantime. The hen pen was still there but another fellow now owned it and he were a bit posh like. Anyway he agreed that we could keep the engine there.

So there it was, in this hen pen and the first job was to make a complete new smokebox. This was where the troubles started. We jacked the front up, took the big casting off and burned away what was left of the smokebox. There's more metal in a bucket full of holes than in that 12 ton of iron. It should have been ⅛in thick but it was paper thin. The owner of the land, the Earl of Bradford was no welder so we had to call on the services of a professional welder.

We got the new smokebox ring rolled round, 2ft 1in wide and 2ft diameter. I then put up three poles and got hold of a set of chain blocks and raised the new component into position. This fellow then came along and welded it in position. The next thing was to put the big casting back onto the front so this too was winched up. This was in the middle of a field so there was nothing like electricity available so I had to drill all the holes with a ratchet drill. The casting was held in place, and a plumb rule was placed on the side, "Yeah, that's right, get the drill out".

And I started drilling the holes and I drilled all the holes, 12 in all. We bolted the thing on and we looked at it again and then put the funnel on. But now things did not look right. The back end of the roller had sunk in the muck and the casting was about ½in out, and the funnel was about 3in out of plumb! It would have looked as though the engine was going sideways.

All the bolts had to come out and the holes then had to be redrilled to make them oval and half moon shaped packing pieces made up and put in the holes to bring everything upright again.

Nearly everyone who buys a steam engine starts at the front and works their way backwards, or they did do when you could buy one that would actually run and more or less just needed cleaning and painting. The rear end is the hard part, where all the muck and the oil is. The smokebox was now on so the next job was to lag the boiler. We made a supreme job of that too, with mahogany lagging and brass bands. The engine rolled around for best part of

12 months with the tin boiler cladding off with the mahogany showing — it looked like Stephenson's *Rocket*, marvellous.

The tender was next and this was just hopeless. There was nothing you could do with it. In the early days we welded patches on to it but the ultimate, end product had to be to build a new one. But making a new one of those! At the time it compared with building the QEII in your back garden. I had a Black & Decker, a ratchet drill, a big hammer and a few spanners. All those rivets and flanged plates. I laid in bed and put my thinking cap on, and thought and thought. How could I reproduce that beautiful flanging that the old timers had done. At last, I cracked it. I made concrete formers with 6in iron nails imbedded in them for reinforcement and had the long straight sides with the radius's put on with a big press. Then I cut them away where it went really fancy and blacksmithed this little piece on at the top at the back, where it is flared out. I had made that as a separate unit and had it welded in position, so really there were only two difficult plates. These were the ones on the sides. The lower part of the back was one sheet of quarter plate that was bent fancily, but nearly anybody can do that on an ordinary press. So I just made a template for that shape and got it bent. I then drilled a "million" holes and did the rivetting. The rivets were held up on the outside and I then bashed them in. When the tank was completed we won a prize for it at the traction engine club as it was something like the best tank they had ever seen.

Then came the problems of the gearing. After 20 miles on the engine you were left stone deaf for about three days. When you got in the Land Rover you could not hear the engine running, it was like a Rolls Royce. The worm and wheel on the steering gear were so worn by this time that they were positively dangerous, because these gears held all the weight of the engine. If you had hit a big stone it would probably have ripped off what was left of the teeth on the worm wheel and the whole thing would have careered off out of control.

The worm wheel had already worn away once before and it had been built up with weld upon weld, the low gears just being a mass of weld. A friend of mine with the same make of engine had actually had a new worm wheel cast as a blank for his, and then got a local firm that makes lifts, to put some teeth on it. They cut worms and wheels all day long every day and I knew a fellow who worked there, so I asked him, when I saw him next,

"Hey, any chance of you putting some teeth on this wheel for me, and making a worm the same as on the engine, or rather what used to be on the engine?"

He replied, "Yeah, yeah. Leave it with me. I know the Managing Director."

So I gave him the old worm and wheel together with the blank wheel my friend had got for me. I came home from work one evening and found the whole lot on the back door step, untouched. So I thought 'so much for him knowing the Managing Director'. I rang him up and asked him what had gone wrong. He said he was sorry but the firm was very busy and suggested I went to a gear cutting firm in Salford. I thought 'they're passing the buck here, to get me out of the road'. Anyway I went to see this man at Salford and had a word with him about my gears; it was like going to see a doctor about an illness. He told me not to worry about it and they would cut the wheel for £30 and make me a new worm for £25.

While I was there I mentioned there were six other gear wheels in the train which were all completely knackered, and I drew a sketch of them on the back of a cigarette packet. He suggested that the best thing I could do was to go back home, get them all out of the engine and take them to him and he would give a price for making a complete new set. So I sweated and toiled to get them out and took them into Salford for his opinion. The trouble was that not one of them was a straightforward gear wheel, each having some sort of complication. It either had bell shaped spokes or it had long splines connected to something else. Also, they have to be cast in steel. If they were cast iron I could have got them made up for next to nowt nearly, but cast steel is a different job altogether, very expensive. The man at Salford took one look at them and said,

"This'll cost you a fortune!"

So much for being told not to worry about it. I noticed at this works that they had a big gear wheel that had all the teeth welded up and they were re-cutting it. I mentioned this so he then suggested that they would get a price for doing the welding and they would re-cut them all for £200. The people that did the welding were in Yorkshire and they wanted £300, just to weld them up. I just hadn't got that sort of money to throw around and told the chap at Salford that it still was not on. He then came back with the suggestion that it might be cheaper to start from scratch, but instead of getting them re-cast, or welded go to a firm and get some blanks run up on a lathe, there being plenty of firms around that could do that. When I had got the blanks I could take them round to him and he could then put the teeth on them — for the £200.

Somehow we did not seem to be getting very far and it was beginning to look as though the whole thing was grinding to a halt, quite literally. Then, one day at about this time I got a job round at the well known engineering firm of Hick Hargraves, whose works I described in the previous book, 'Steeplejack'. Whilst I was there I so happened to mention, in conversation sort of thing that I had problems with my gears. Now this kind of job is nothing to a firm like that and the fellow said, as I have so often heard before

"Leave it with me, I will see what I can do. . ."

With nothing to lose I took all the old gears round to them and in no time

*"Alison"—the steam roller, as now fully restored in lined green livery and known to millions of television viewers. Depicted at the May Day Steam Rally, Abbotsfield Park, Flixton, Manchester on 7th May 1984.*

9

at all they had machined all the blanks and put the teeth on, all for an unbelievably low price! I had been trying all over the place to get the job done at a realistic price, and it ended up here in Bolton. What a difference the new gears made. They called it the electric steam roller after that it was so smooth. All the vibrations had been cut down, together with the racket. Before, the little lids on the oilers used to shake up and down and when you put your hand on the framing you could really feel the punishment it were getting. All that just disappeared overnight. It was now beautifully sweet and in low gear it sounded more like an electric tramcar when you opened it up, not all that clanging and banging.

*Detail of a couple of the plates that adorn Fred and Alison's steam roller. The Aveling & Porter builders plate is in effect the engines pedigree.*

At long last the gearing was now perfect and to cap it all off we made a fine new canopy roof and gave the whole engine a beautiful coat of green paint and nicely lined out in black, gold and red. The brass nameplate *Alison* was proudly fixed to a job well done. It had taken 14-15 years to do all this work and after all the time, money, effort and energy spent on the thing, it was now complete and something of which we felt we could be justly proud. In fact this was recognised by others because at the next event we attended we won the cup for the best steam engine. Later that week the firebox popped.

*Close up of the engine showing the brass nameplate, the builders plate and the original owner's plate — "Flintshire County Council No. 3". The rivetting on the front of the firebox, right, is also clearly visible. This called for 232 holes to be drilled by hand using a ratchet drill.*

# Restoration and Frustration

I had always known that there was more welding in the firebox than in the QEII and had said that as soon as it sprang a leak I would pull the engine to bits and make a complete new boiler and firebox assembly. Well, it had now sprung a leak and we had it bodged up temporarily by a welder.

*The delicate touch needed to keep the engine on course is demonstrated by Fred*

We ordered the new boiler plate which duly arrived and this is where the troubles really started. If you undertake to build a boiler you have to have a number stamped on the pieces of plate you are going to use, and you also have to have a test certificate. This is the British Standard Government Test Certificate which specifies the quality of the plate, how much carbon there is in it and how many tensile stress tests pieces there are and how many tons it will hold and all this jargon, and so on.

Once the boiler test insurance company has satisfied itself that this plate is suitable you are then given permission to start the bendings, i.e., rolling the piece round for the boiler barrel. Well, we had got the O.K. and so we rolled it round for the boiler and did the appropriate weld preparations. The inspectors came along, had a look at it and said it has to have a 3 mm gap, and it has got to be this and it has got to be that. Again, they O.K'd all of that and we got a firm to weld it up. These people seemed to know what they were doing, after all they did work for North See oil rigs, British Gas, made pressure vessels, the lot.

I then had to take it for a tensile test and bending job on the two plates, which are welded to the longitudinal seams. It was X-rayed while I was there and they said,

"It's lovely, there's no cooling cracks, it's very nice. Take it away."

So I took it home and drilled about 90 ¼in rivet holes when another boiler inspector rolled up and he said,

"I wouldn't go any further if I was you. The elongation on the test specimen of the weld metal is 3 decimal points short of what is allowable. You will have to contact Glasgow to see if it is acceptable".

*whilst below, he shows that more physical effort is needed in order to steer the 12 ton monster round a tight bend!*

It almost seemed to go without saying, as of course they did not accept it. The people who welded it offered to do it again for no additional charge, and they even had the findings of this analytical chemist chap who worked for the gas board. He had said that there was nothing wrong with it as it was so near the mark that it could have done. The rules are so 'over the top' that there was no way it could ever have 'popped'. The firm told me to take it back and they would grind all the weld out and they would weld it up again. The drawback with this was that I would have to pay another £70 for a further set of tests on the finished job. This was a lot of money for something that still could not be guaranteed, not that I had that sort of cash spare at the time anyway.

I therefore asked the insurance company what they would do if I presented them with a fully rivetted boiler. There are no such tests for that. Providing all the centres of the rivet holes are drilled in the right place for that thickness of plate, a double rivetted seam is put in and the holes are clean, good and fair, that is O.K. In other words if the rivetting looks visibly attractive and there is not one up here and one down there, it is an acceptable join. The only actual examination is the hydraulic test and that is at the end of the line. The reply came back from Glasgow, something along the following lines,

"Dear Mr.Dibnah, Where in this day and age are you going to find someone to do the double rivetting lap joints for you, if you do decide to use this somewhat outdated procedure?. . ."

I thought, 'well, if I cannot drill 50 odd holes in two straight lines, with the right centres, I'll eat my oily cap!'

Here again though we were back to building a completely new boiler. The other thing had already cost seventy odd quid just for the plate, plus the welding and the tests and all we had achieved was turning a perfectly good piece of plate into a load of scrap. We got £16 for it at the local scrap yard.

Back to the drawing board. This time we acquired a length of large diameter gas pipe as I had actually known someone who had made a boiler out of this stuff. The pipe was slightly too big for our boiler so the plan was to cut a slot in it and then tighten it up, to end up with a four inch lap. The insurance company in Scotland demanded a piece 12in square so that they could do the usual tests on it to make sure that it was suitable boiler material.

The inspector down here made the good suggestion that before we sent this up that we should carry out our own tests, these being quite straightforward. We looked up some old boiler making books and found out what had to be done. The British Standard test piece for boiler plates had to be ½in thick and was a strip 8in long and 2in wide. You get this strip, put it in a forge and get it red hot and then bend it back on itself. If there were no flaws on the outer circumference of the lap then it is acceptable.

You then bend another piece, the same, ½in thick, 8in long and 2in wide, but this time it has to be cold. It is permissable to leave a gap between the two ends, something like 1½ times the thickness of the plate itself, as you would probably find it impossible to do otherwise.

Therefore we cut a couple of strips off this piece of gas pipe, put one in the fire and got it red hot and bent it over, beautiful. Not a sign of a crack or anything — beginning to look quite hopeful.

So we put the next piece in the vice and hit it with a sledge hammer. It were like a piece of spring steel — the hammer went "boing" and nothing happened. I simply could not bend it! Now, a normal piece of boiler plate is as soft as cheese, it bends that easy. Not to be deterred I went on and did bend it but what a job, it nearly broke the vice off the bench.

With those two tests achieved we could then go on to the next, so we cut a third strip, ½in thick, 8in long and 2in wide. For this test you put the piece in the forge, get it red hot and then plunge it into cold water. Then you try to bend it. The particular piece that we did this test with had been drawn to a fine feather edge. I got it red hot and plunged it in the water, put it in the vice and hit it with the sledge hammer. It broke clean off like tool steel with a resounding "ping", shot through the air like a dagger and impaled itself in a tree! Even I was not going to make a boiler out of this stuff. So that were another load of scrap.

I finally found a real boiler works where they would roll round the piece of plate for £50. It was a British Steel Corporation place in Yorkshire, a very old place and they put this beautiful stamp on it, complete with a crown, something no boiler insurance company could argue with. Also it came complete with a test certificate just to make sure. The plate was rolled and was returned with the edges planed and ready for caulking. A superb job from a proper boiler works.

With confidence restored I now started to drill the holes with everything now going well. When you have never made a boiler before, it can be a bit frightening when you look at what is left of it when you have finally got it all stripped down and the bad bits discarded. All we now had was the outer shell of the firebox wrapper with no boiler barrel and with the stays out it were like a great big cullender, as you could see daylight through the sides.

The back plate, (the firehole door plate) was badly wasted along the bottom so we had to cut a great big piece out of that. So we were now down to just the two sides and a top of the firebox from the original boiler assembly.

One day, completely out of the blue this boiler inspector turned up. I had not realised this would happen like this, and from time to time he just popped in whilst he was passing, usually before I went to work at 9.00 am. He came in, had a look at the progress from last time and would go away again. Then about a fortnight later I would receive a letter,

"Dear Mr. Dibnah, 53 rivet holes examined today and found to be all in order. Yours faithfully . . ."

This went on for the whole of the 18 months it took me to complete the boiler, drilling and rivetting at weekends and in the evenings. Finally it was all finished and it was time to put the water in to see if there were any leaks. We had got our own hydraulic test pump and we pumped the thing up and it were quite unbelievable. Leaks appeared all over the place. These spurts of water were as thin as a spider's web. You did not need a big hammer to stop them. A 'dub dub dub' and it was gone from there. Another 10 pounds of pressure was put on and it came out from somewhere else.

You can spend hours stopping these endless leaks. You think you have stopped them all and then all of a sudden you would see another one, not gushing out, but a jet, smaller than the smallest water pistol. The squirts of water are like fine silver threads and if you put your hand there it is wet through in a second or two.

Finally we were there with the working test pressure, up to 200 lbs p.s.i. and all leaks stopped except from one rivet in the firebox which simply could not be got at. It were right at the top and when the boiler inspector arrived later in the day I asked him what could be done about it. It took me by surprise when he suggested I just ran some weld round it. But I thought, to myself, 'I've spent all this time making this boiler in the old fashioned way to get it right and no way am I now going to finish it all with a dollop of weld!'. I was determined to stop the so and so somehow or other.

He then said the only other way he knew of was to empty the boiler, let it rust up a bit, then fill it again and try it. This would take months and I wanted to be on the road during the next week or so. I laid in bed, thinking cap on again. This rivet was only 2in off the top of the firebox so you could not get at it. Your knuckles were knocked red raw trying to go round the corner. You could get ⅞th the way round but it was impossible to get onto the top. I then came up with the idea that if I got a 3in x ½in steel bar and bored a 2in hole in it, ½in from the top and then put that over the rivet head with two big bolts through the firehole door, and a plate on the outside, I could squeeze with the caulking tool on the edge of the 2in hole and onto the nib of the rivet.

It worked, it stopped it and the hydraulic test was completed satisfactorily. Then the next thing is in my opinion something a lot of people have come unstuck with. They have sent a boiler to a boiler shop where it has been repaired and given a hydraulic test. This is of course all done cold — cold water, cold iron, everything. One of these boilers expands in steam to the extent that it is about ⅜in longer than when cold. I have heard of several people that have received the boiler back from the shop, complete with a hydraulic test, then cannot get the engine back together again quick enough. They have put the tender back on, got it all finished, lit the fire and then it has sprung a leak on the foundation ring. This is right in the worst place where you cannot get at it without pulling it all apart again. Then they threaten to sue the boiler company and all of that, for doing a wrong job.

*The prancing horse motif "Invicta" of Aveling & Porter Ltd of Rochester and Canterbury, Kent. This was one of the most prolific builders of traction engines and steam rollers and consequently one of the best known throughout the world.*

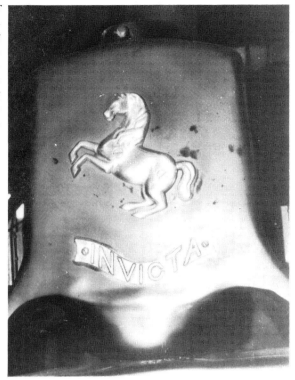

Instead, you should light a fire in it before you get the rear end on. Therefore I decided to make sure of it by steaming it about fifteen times before we put the tender back on. In the meantime I received a letter from the boiler insurance company:

"Dear Mr. Dibnah. We understand from our inspector that the work on your locomotive type boiler has come to a standstill. Here is our fee for the supervision of the work so far. £175".

£175? They had done nowt, sort of thing. I had expected a bill of something like £30. Apparently everytime the inspector had come round, which was when ever he felt like it, and sent me one of their papers saying how many rivet holes he had counted, it had cost me twenty quid. He had not even put his overalls on at any time. I thought it was time to get another firm.

This is where it starts to get funny. I was working at a well known tailoring firm at the time and they had just had 20 odd rivets put in a Lancashire boiler. This bloke had charged them £1,500 for the work — and it leaked. You could actually get your fingernails behind the heads of some of the rivets where you could not see them. So I asked the engineer what the boiler inspector had said about it.

"It'll seize up, it'll be all right", is all he had said apparently.

So I said, "I want a boiler inspector like that! What's his name?"

He told me that it were a mister suchabody like, and gave me his 'phone number. I rang him up and instead of just saying I had got a locomotive boiler I would like him to look at, I told him the whole tale, from start to finish, but he still seemed interested. All he asked in fact was,

"How on earth do you build a boiler in your back yard?"

He then said he would come down the next Wednesday but in the meantime he would have to inform his superintendent in Manchester and let him know what he was about. Wednesday morning came round and we were waiting, in great anticipation. At 11 o'clock precisely the 'phone rang.

"Hello Mr. Dibnah. It's mister suchabody here. I have told the superintendent where I was going and he said — "Stay away from it! Don't go near it — it might blow up!"."

We were now in trouble — we had about a week and I wanted to be on the road. I knew the thing would not blow up. We had had umpteen hydraulic tests on it and it looked beautiful. In fact had you have left it out in the rain and let it go a bit rusty you would never have known it from the original. There was not a spot of weld anywhere, it was all perfectly rivetted.

As a last resort I contacted the people who do the insurance on my Land Rover, and they said,

"We don't do things like that. We only do houses and cars normally, but as you are a customer of ours, leave it with us and we'll come back to you".

How many times have you heard that one? Then, much to my surprise, it was about 6 o'clock one evening and the 'phone rang. It was the fellow from the insurance company and he said he would be coming round to see the boiler next Wednesday if that was all right. We were nearing desperation by this time so I was prepared to agree to anytime of the day, or night. Anyway, there we were again, on the Wednesday anticipating the arrival of the insurance inspector. So I said to Donald, (it happened to be raining that day so we could not work outside),

"Steam the thing up and we will have it on the drive and blowing off when he comes round the corner. That should impress him".

It did look impressive too, there was steam everywhere. He duly arrived, in his £200 suit and a big fancy car and walked up to the engine and looked at it and then stood back, with obvious wonderment in his eyes.

"Have you made this here?" he said.

"Yeah, we made it here all right, over in that shed", I replied.

He then walked all round it, inspecting it very closely, here and there and

*Fred stops for a chat and a smoke between appearances in the arena with his engine at the annual Flixton event which is jointly organised by the Urmston & District Model Engineering Society Ltd and the Lancashire Traction Engine Club Ltd.*

then stated,

"It's marvellous. I can't see anything wrong with it at all, anywhere".

This was more like it. Donald and I looked at each other, and smiled. Then he said,

"But I'm not a boiler inspector . . ."

I was then told a mister so and so would be coming to see me. I had already heard about this fellow from some of the other boiler lads at a firm in Bolton. They reckoned he was really hard and went absolutely by the book and no messing about. The inspector from the previous firm was easy meat compared with him, in fact he had been a decent lad really and we had been on first name terms. It were the company that had caused the agro.

Eventually this fellow rings up and stated bluntly,

"I'll be coming next Thursday. Make sure you have everything ready for me".

This time we had the hydraulic test pump on and this man did all sorts of things to the boiler that none of the others had done. He asked for a piece of 3/16in iron rod and wanted a needle point ground on to each end. He bent this round so that it touched the crown of the firebox in the centre and it was about 4in above the foundation ring at the bottom. He put a centre pock mark at the bottom and one at the top of the firebox and put it on these. When I pumped it up he called me over and said,

"Look at that".

It was amazing. You could actually see the pressure squeezing the firebox, stretching the crown bolts. They were moving about a good ⅛in when the full pressure was inside, and that was cold. It would be even more when it is hot. He looked at it again and said,

"Right, we'll open it up and let the water out".

We then opened the big door on the side and he peered inside and shone his lamp around. They don't say a lot, these men. He knocked one of the mudholes in at the bottom and put his fingers in there and had another look round with his torch. He asked if it was like that all the way round and when I said it was he did not bother about the other two.

He seemed quite happy with it all but then turned and said,

"I'm sorry I can't come back this week. Next week it's the annual holidays and so we are very busy going round the cotton mills. In fact it will be three or four weeks at least before I can come back to see it in steam".

Three or four weeks? I had a different idea, and knowing he lived in Bolton suggested the following as an alternative.

"You've only got to see it blowing off and have a look round for leaks haven't you?"

"Yeah, yeah".

"Well, can't you come back this evening, after tea? I can easily have it in steam for then".

"Well, all right then," he agreed reluctantly. "Can I bring the lad along too, he's only a kid like".

The two of them came back early in the evening and when they arrived it was already out in the yard, blowing off. I suggested to him, just to prove it was not held together with chewing gum that we should take it and go down into the town for a ride over the cobble stones. So there we were at last, taking it out on a tour of Bolton. When we got back home he said,

"It's a credit to you. Your ticket will be in the post in the morning".

When the boiler report did come it was a very glowing one and read,

"This boiler has been extensively renovated in so much that it has had a new boiler barrel, a new front tubeplate, a new firebox, a new outer wrapper for the firehole blow plate, and the platework and the rivetting are of the very highest standard".

This is very good as they do not commit themselves as a rule, just in case something does go wrong.

Really we had done this job the wrong way round — building the boiler last. Even today it has the same paint job as it had before. We managed to pull it all to bits and put it back together again without damaging it. But somehow, mysteriously, we have lost one red line on the boiler, the cladding must have got shorter over the years.

*The steam roller is seen at rest, showing the flywheel side.*

## Chapter 2

# Starting all over again. . .

People who restore a steam engine are nearly always heard to say at some stage before they complete it,

"If I knew what was involved in the first place, I would never have started, and certainly never again . . ."

Then, almost immediately it is finished, they start talking about the next engine! It's a bug that once it has got you there's no escaping.

So it was with us as we have now gone and got another one, but at least we know a bit more about making boilers this time. The story of this one actually goes back about 17 years to when we bought the roller. A friend of mine had acquired a tractor, or more correctly a convertible, as it was originally built as a tractor but had been subsequently made into a steam roller.

He had given £200 for it and had brought it up from Devon County Council. It was another Aveling and Porter, No. 7838 also built in 1912, and it was beautiful compared with what we had just spent £175 on. Ours was a clapped out heap, this was actually a tractor to boot and it was a compound engine and even the paintwork was in fair nick. On the other hand though, it had no front axle as this had been left down in Devon as it was intended to restore it to tractor status, this being a great advantage when attending rallies and other events under your own power.

For 16 years he had done nowt with it. When he moved, he moved the thing with him and it got carted around all over the place. Eventually he decided he would pull all the lagging off the boiler and have a look at it to see what it was like. The lagging was taken off, and what a mess! There was a great big patch on the side of the boiler which had once been ⅛in thick. It was now so thin that I put a cold chisel right through it with no problem. There was no way this boiler was ever going to be steamed again. New boilers are an expensive business and the owner was not in a position to build one himself so rumour had it that he wanted to sell it.

I thought a £1,000 would get it and that would be a fair return on his original investment, especially as nothing had been done to the engine over the years. I had a spare couple of grand at the time and my idea was that a £1,000 would have bought it and the other £1,000 would have paid for the boiler plate and the rolling for the new boiler barrel, together with the new tubeplates and so on. All I would have to do was, as before, drill a million holes for the rivets.

When the opportunity arose I had a word with him about it but he kept jacking up the price. This got up to about £1,500 and I thought I don't want to know no more, after all, what I would be doing was buying myself about five years hard labour. So we left it at that, for a while.

I continued to think about it, but realised that I did not really want the thing anyway. It were mainly the wife's idea, just to keep me out of the pub. I argued that having got to £1,500 there would be nowt left to buy all the bits and pieces that were needed.

A few weeks later we were at a traction engine rally and Alison said to Peter's wife, that being the bloke who owned the engine,

"Why don't you get Peter to sell Fred that engine?"

I had had a few pints by then and muttered something like,

"I'll give you £2,000 for it".

But I thought to myself,

'You fool, you're drunk!'

I was quite relieved when he said he wanted just a little bit more for it. We had been joined in our conversation by one of the chaps from the BBC who were making a film about our appearance at the rally. About a fortnight later we were talking to the television director, who had been told that I was interested in buying another engine, and he said,

"You really should buy that engine Fred. Then we would be able to make a film of you restoring it".

That sounded like the easy bit to me, but he kind of persuaded me that it was an opportunity I should not let pass by. Not long after that we bought the engine and moved it into our yard where I started work, and they started filming. We made good progress to start with, and built a new boiler, a new front axle and all the front bit and I drilled all the rivets out of the foundation ring. I had, what I thought at the time was a bit of luck with one of the normally more expensive items required, namely the front tubeplate. I was wondering how I could get this done, at a reasonable cost when I had a 'phone call from a firm who wanted a small factory chimney bringing down. The chimney itself would have presented no problems in itself, but it was situated at the end of their workshop building which had a glass roof, so this meant it had to be brought down a brick at a time, like they had seen me do it on television.

This was at Blackpool, so I said this was a bit too far out of my usual area, especially for what sounded like rather a smallish job. So the chap replied,

"It's a good opportunity for you to bring the family along and have a holiday by the sea at the same time Fred".

This did not sound too much of an incentive to me, after all Alison and I had already had a holiday, when we had got married. But then he said,

"We make pressure vessels here and I also thought we might be able to come to some sort of an arrangement like, if you needed anything".

'This is more like it' I thought to myself and told him I was in need of this tubeplate for my engine.

We all went off to sunny Blackpool by the sea, Alison, the three girls, myself and of course the BBC television people who were still following us around at that time. The first few days were terrible, we spent all our time on the beach. It were warm and sunny, but there was more wind there than at the top of 200 ft chimney. Then we had a really wet day, definitely not beach weather I said, so I got Alison to help me put the ladders up the chimney. I don't think she was too happy about it though, not that she said much at the time, nor all of the next day for that matter.

Anyway, while I was knocking the bricks off and dropping them down into a skip below, having a marvellous time, the firm were making the tubeplate for me. This was going on directly underneath where I was working. A lovely job they made of it too. In fact it all seemed to work out very well in the end. I got my tubeplate, the female Dibnah's got their holiday on the beach, the firm had their chimney removed and the BBC got their film. This episode has now been shown on TV at least three times that I know of and has proved to be very popular viewing — with the taxman if nobody else! They promptly made a video of it and then came round and assessed me for all the work done and paid for 'in kind' as they put it. That is what they estimated, going back over x number of years, and that is a whole story in itself. The end result was that my tubeplate, instead of being a reward for my labours turned out to be a very expensive item. . .

Following the completion of the television series life has got a bit more complicated in lots of ways, plus I am building another half on to the side of

*This is another product of Messrs Aveling & Porter, their number 7838, built in December 1912 and is an 8 ton machine. The registration carried is TA 2436. The all-new smokebox is shown here.*

the house. This is something Alison has been on at me about for some time. The kitchen is only 3ft wide so she does not have enough room to pull the drawers open. I promised to make a start on building the extension after we had brought the chimney down at Leigh, at the end of 1983 as I thought that would provide plenty of bricks. In fact Alison was heard to say on a radio broadcast of the event, as she looked at the pile of smouldering bricks that had been a 200ft chimney a couple of minutes before,

"That's my kitchen lying there!"

*Another view of the as yet un-named engine, showing the new chimney.*

But they turned out to be poor quality brick and they had all been smashed to bits, so she had to wait a bit longer, until the next chimney job.

Therefore I have had to stop work temporarily on this engine because of more pressing needs you might say, but I have got several parts in works that are being done, as and when the people have the time. I recently took delivery of a new pair of rims for the driving wheels so it is still progressing and I hope to get back on to it in the winter and get a bit more done. I would not like to say when it will be completed, but it will be one fine day.

With all our other engagements and activities these days we do not even get out so much with the steam roller as we used to. This year all we managed was our annual outing to the May Day Steam Rally at Abbotsfield Park, Flixton, Manchester but at least we did get an award which is a cup inscribed: "The Ernie Cooper Memorial Trophy L.T.E.C. Members Best Steam Roller". The only other event we attended during the year with our own engine was a carnival at Westhoughton.

*New rims have been obtained for the rear wheels and were photographed in the yard in October 1984, ready for the centres to be fitted.*

*The boiler and firebox assembly is depicted here as well as some of the gear wheels. The rear, driving wheels have still to be fitted.*

*The Dibnah's second engine was acquired as a heap of bits and pieces, but after much work is now recognisable again as a traction engine. This has road wheels, rather than rollers and will therefore be able to travel at much greater speeds than the other engine.*

# The Living Van

When we do go out with the steam roller we always take the living van with us as this provides us with accommodation as well as extra space for coal, this being carried underneath the beds. This vehicle is a story in itself.

These vans were always sent with the steam roller and the driver, all over the country, being conveyed to their destination by train. The driver lived in this van when he was away from home, and there were strict rules about this and later in the book we have some extracts from the driver's instruction book, including the line:

"Under special circumstances no objection will be taken to a man's wife occupying the van for short periods". The driver had his wages sent to him through the post office local to where he was working.

When the use of steam rollers came to an end they did not bother to have the living vans taken back to the depot, and they were just left laying about all over the place. I was told about one up at Hetton-le-Hole in County Durham so we went up there to see it with the idea of buying it, and bringing it back with us. We would restore it for use with our engine. When we got there we found it was a genuine effort and the fellow who owned it was keen to sell it. It had been used as a hut and did not even have any wheels on it and he was asking £90 for it. Ninety quid for what was no more than a rotten box, so we just left it at that and turned round and headed for home again.

We decided to travel back down the country lanes rather than zoom down the motorway. We were coming along a road, up a steep hill between Clitheroe and Burnley when we saw in the middle of a field full of sheep, what we thought was one of these living vans that we were after. So we stopped and had a look, and it was one, just standing there in the middle of the moors. It was painted bright red and was about 200 yards from the road,

I went along to the farmer's house and knocked on the door. A young girl answered and told me the farmer was out, so I said to her,

"Tell the farmer I'll give him £30 for that derelict hut on wheels in the field, if he wants to sell it".

"Oh, I expect he would", she replied, "as he's only going to burn it at the end of the week and recover any scrap metal from it".

We exchanged telephone numbers and we continued on our way home to await his call. By 11 o'clock that evening we still had not heard from him but half an hour later the 'phone rang and it was him.

"Give us £35 and it's yours" the farmer suggested.

Next weekend we went back up there — a lot nearer than Hetton-le-Hole as well, and collected it. It did not have a drawbar so I had made one up and took it with us. We brought it back behind the Land Rover and it certainly slowed us down when going up hills, it was squeeking and groaning all the way.

Once we had got it back home we were able to have a closer look at it and began to realise just why the farmer had decided to burn it. It was in a terrible state! Some things were obvious from the start, like one side had been cut out and a picture window from a house had been put in, and there had been a fire in the corner where the stove was. The door had been replaced at some stage, the new one appearing to have come from a toilet; well it still had a toilet roll holder attached to it. The ceiling was all cracked up and looked like crocodile's skin. This had been whitewashed over when the farmer's wife had decorated the inside with purple and orange wallpaper. The charred roof had been covered on the top with metal from old oil drums and tarred felt.

Underneath the van we discovered that the main oak beam and the

*The living van which is hauled by the steam roller whenever the family go on an outing. In addition to accommodating Fred and his wife Alison, the three daughters, Jayne, Lorna and Caroline, plus the Alsation dog Bonnie, it also acts as an additional coal tender. Enough fuel is carried for the outward journey at least, in sacks under the beds. It also serves as 'camp' at the event as seen here at Flixton in May 1984 with additional bags of coal alongside. It was in plain green livery at this time having just been repainted after repairs necessitated by its argument with a passing van driver one night.*

longitudinal runner along one side were rotted away. Also, four of the cross members which looked perfect on their ends had their middles completely rotted away. These would all need replacing. It stood on very vintage looking wagon wheels probably dating from the 1920's, very big and narrow and not the right type at all. However, I did already have a set of the correct pattern, cast iron wheels which would be used.

You probably would not realise it, but a thousand feet of tongue and groove board goes into making up the sides of a van like this, and it nearly all needed replacing. I got various quotes from timber yards, ranging from £120 to £150 for this item alone, I dare not think about that oak beam etc. Then I got one more quote and this came back at £60 for the t. and g. so I went along and selected enough material to do the job.

We then went along to the timber yard and got these great big elm planks which still had leaves growing on them. These were taken along to my brother-in-law's sawmill and he cut the stringers out for me. The timber was wet through and you could not use it then as it takes an inch a year for timber to season properly. These were two inches thick so we put them on some skids and piled on as many lumps of old iron as we could and left them there for the next twelve months.

Then, in the summer we ripped all the top off the van so that we could work on the underframe. These timbers went into oak runners. 3in square, fitting into slots. We put them in there and belted some 4in nails in and then reboarded the top, using 2½in nails that nearly went right through these things, so there was no way they could twist or bend. Sheet iron was then put across the top of it all and this was screwed down. We did not varnish the stringers but left them to the atmosphere for another three years and now they have been varnished they are all right and they have never bent, twisted or shrunk.

It still was not finished because on the full carriage they would have had all chains and things for the steering. Also, the front were all knackered and I had to make a new one with mortise and tenon joints with new iron shackles around it to keep it in position.

Then the back end were bit of a nightmare. The piece of timber that the back axle was fixed to had hundreds of great big rusty nails knocked into it. These had been banged in there for carrying oil cans and tarmac rakes and all sorts. They were all bent and broken, a horrible mess. This were an enormous piece of oak 8in square and I kept looking at it and putting it off.

This was until Greenall's came along and wanted to have the van painted in their green livery and carrying the inscription "Greenall's Local Bitter", as a kind of travelling advertisement. I pointed out that they could not really do this super paint job with the back axle falling out, as I thought this was the opportunity to get rid of this thing at last. It seemed a good idea to let them pay for a coach building firm to carry out the work. I got various quotes in and they all wanted a fortune — I was obviously in the wrong business. This was for just doing the axle so I said to Greenall's,

"Give us the wood and I'll do the job myself like".

I took four days off work and did it with a huge piece of oak they brought along. It was the correct 8in square, but I don't know who measured its length, because they paid a lot of money for it and we ended up with an off-cut over 5ft in length. For the two runners that go across the back, to take springs they sent along solid mahogany for them so I did this bit myself too.

*A detail view underneath the rear of the van showing the axle and leaf spring. An oil lamp and a can hang from a bracket on one of the replacement timbers which were fitted by Fred during its extensive restoration.*

Finally I made a tool box for the back so by then the whole thing was more or less brand new, apart from the frame-work which still had about ⅛th of the original left which was all right.

To build something like this now at modern day prices would not give you much change out of about five grand I would imagine, and that would be just for the wood.

It were stood outside on road, where we used to park it beside the house, earlier this year and the police were chasing some midnight maniac. As he went by he crashed into the side of it, he were driving a van at high speed and it ripped all the side off his vehicle. All the damage it did to our wooden van was to one board on the back and two on the side, but all the pop rivets had shot out the aluminium sides of his van and were sprayed along the side of ours like pellets fired from a shot gun. The paintwork were full of all these little dinges along one side. After the repairs has been made it was decided to repaint it, and as the arrangement with the brewery had come to an end some time before, it was finished in green again but now has our own inscription on the sides and back.

Since that incident we have not kept it out on the road but take it down the back, the main difficulty being in turning it round to get it out of the garden the next time we go out with it. The engine is kept in the shed which is then on the wrong side of it, so what I could really do with now is a turntable. . .

*The rear of the vehicle as now elaborately lettered-(the modern "Long Vehicle" sign looking rather incongruous!). This van is believed to have been built originally by John Allan & Ford of Oxford, early this century.*

*Chapter 3*

# Stationary Steam

If you have got any type of steam engine you have to keep the thing covered up, somehow or other. When we first came to this house, just being able to have the engine standing outside the back door were a step up from where we had it before. There you had to have it sheeted all over, right down to the ground and ropes tied all round it to stop the vandals from getting at it. Every time you wanted to do something on it you had to undo all the knots and lift the heavy sheet up and over the engine, then get all the tools out before you could even start work. When you had finished work it had to be sheeted over again and all the knots tied up again. We seemed to spend longer messing about with all this tackle than actually working on the thing.

At the house where it is nice and private you could at least just leave the sheet tied to the trees so that it formed a sort of mini shed over the top of it. The trouble was, as before, it would get full of dead leaves, rain water and all sorts of things. When you go and pull the sheet off you get drenched with bath tubs full of stinking, stagnant water. Not very nice.

We decided to at least put a roof over the top of the engine so we got six telegraph poles and dug six holes in the floor, to the same spacing as half a dozen roof trusses I had got from a secondhand woodman once before. After we had sunk these in and filled in the holes all round them we put two more telegraph poles along the top of these to give us two barn type supports. The roof trusses were then put up but we could not afford any boards to go on top so we nailed 3ft x 2in spars across and put the sheet back on top of this. It was still only a shed with a rag roof but at least it was up there and better supported than previously and allowed me to work unhindered underneath.

Then I had something of a windfall you might say. They were building a motorway nearby and one night I met the contractor in the pub. We got chatting so I asked him,

"All them 8ft x 4ft sheets you used for the concrete shuttering. How much a piece are you selling them for — must cost you more to move them on to the next site than get rid of them surely?"

"£20 would get you twenty", he replied.

That is how we came to get the roof boarded with marine ply but I still could not afford any proper felt so we just put up the cheapest roofing felt we could get. Might be all right for a potting shed and it lasted maybe two or three years on our big roof, then it ripped up in the wind and the rain proceeded to drip in. Then we got the ultimate in roof coverings. Ex British

Railways plastic sheets — they will last forever them. We covered the roof with these and it were like that right up until last year when we did this television advertisement for Redland Roofing Tiles.

We were on location as they say, and so I said to this man from Redlands, "Have you got very many of these tiles, sort of thing?"

"We've got millions of them", he said proudly.

"Oh, good", I commented, "because I only need a few, just to do the roof on my engine shed".

A few days later I came home from work and found a lorry had been along and off loaded all these tiles outside the house. We had to work well into the night just to get them in off the road. The shed now has a superb roof absolutely ideal for when the east wind is howling around outside.

The sides of the shed presented other problems but there is so much wholesale demolition going on around here, with factory building window frames just being smashed to pieces, glass and all. Some of them are nearly brand new replacement windows that were put into the building just before it was condemned. They knock these down just the same. If you can get there in time and the demolition man is a benevolent sort he will say,

"If you can get them out before we knock the wall down, you can have 'em".

Therefore we managed to get enough windows to do all the front and the back for a start. But what do you do to fill in all the spaces between the windows? Another windfall came our way. We met this man who electrifies church organs with silicon chips. He told us that all the big organ pipes are then made redundant. These are the great big pipes, right at the back of the organ chamber, that people in church cannot see anyway, as they are out of sight. They would not like to have the other pipes they can see removed because people like to have a church organ looking like one, but it does not matter to them if the other pipes are replaced by technology.

The biggest of these pipes are 18ft long and 2ft square, going down to 6in square and which are not worth having. But the others, they are 1½in thick and made of yellow pine. We bought the whole set of organ pipes for twenty quid and when they were cut down one pipe gave us four boards, 18ft long and 18in wide. These cover a fair area and that is what the boards are on the side of the shed. You could not even go and buy pine boards as wide as that today if you wanted to. The floor is made of railway sleepers which are nice for scrawping about on, much better than scrawping in the dirt like when we first come here.

I have got to build another shed now. This will be the boiler house, to go over the top of the new boiler and the 3ft circular saw. It is murder over there in the winter when it is snowing hard and all you have is a roof over your head with no sides on it. Anyway I have got a set of window frames for this which I discovered in the joiners shop of a mill that was being demolished. They were brand new and were going to be smashed up until I put my spoke in. Cost me £50 just to put the glass in them though.

The great dream with steam is to have the whole workshop facility self sufficient. I know it would make life a lot simpler to have a little 10HP electric motor purring away but it is the electric bills you have to think about. The meter box would have its dial going round like a 78rpm gramophone record the amount I use it. Whereas the steam engine almost literally runs for nothing. It will run on sticks and rubbish and a bit of coal, and there is plenty of that around here even without the services of Mr. Scargill and his crew.

When I first got some power tools going we drove it all with the steam roller. That were a bit noisy though because the transmission had to go round a 90° bend. The crank shaft were at 90° to the main shaft so you had to get the power round the corner. There are two ways you can do that, one was to go round the bend with a belt on two pulleys in the roof of the shed. This would have been very complicated as it would have meant having to line it all up again each time the roller was moved.

We managed to get hold of two great big cast iron bevel wheels together with a gearbox and stuck them in the shed. After that all we had to do was the simple matter of putting one belt across the main shaft and another off the flywheel onto the pulley at the end of this gearbox. But that made an unbelievable row. You could hear it half a mile away. So to deaden the sound we put it all in an iron box full of sawdust and oil which made it sound like a Rolls Royce.

Local people know about me having the steam engine and they bring me bits of wood for fuel. A lot of people these days get stuck with an old gate or a broken window frame and ask me to get rid of it for them, because nobody has fireplaces any more. Some come along and throw the wood over the fence and down the banking beside the house. This meant that when my workshop was driven by the steam roller I had to walk all the way across the garden, pick up a lump of wood, walk all the way back with it, through the engine shed and saw it into bits on the circular saw, which were connected to the shaft. Then I had to walk back through the shed with it and put it in the coal bunker on the engine so that it could be fired up. I spent more time walking about than working.

It struck me at the time that I could extend the shaft right across the garden as there is very little friction in turning that round. I could put the engine, and later the stationary engine, which at that moment in time was not renovated, at that end of the garden which was the fuel delivery end of the whole set up. Also at that end would be the boiler that would make the steam to drive the engine to turn the shaft to work the machinery in the shed.

I then proceeded to extend the shaft out of the shed and this were no easy

*General view of the garden/yard in autumn 1984, looking towards the engine shed/workshop. The shafting can be seen running right across the area, through the trees.*

feat. Anyone looking at the thing would think to themselves, ' Well that's dead simple isn't it'? It's just a big long iron bar 98ft long going across a garden on sticks'.

But you want to try and do it! And make it go at 80 revolutions a minute without shaking the place down. It is a bit more difficult than you might think.

The first part of the job were to span a distance of about 25ft which could not have any supports underneath, because they would have been in the way of the engine going in and out of the shed. We put the first trestle up made of telegraph poles, then we raised up a piece of shafting and looked at it. At 25ft long there were about an inch and a half of dwell in it and that is no good for anything. We then took that down again and decided that if we got a great big piece of heavy gauge tubing, the right shrink fit onto the shafting we could get one end red hot in the forge and shove a 3ft length of shafting up it. When it cools off, "boom" it has the grip of death on it and so no way will it come adrift in the air. We did this at both ends of the tubing and raised that up into position.

When we looked at that it were perfectly straight this time over the whole 25ft, so we put a coupling on one end and connected it to the shaft in the shed. We put a bearing on the end and packed it up and lined it all up and proceeded to raise steam in the steam roller. The shaft started to go round but after only a few revolutions it did not seem quite right somehow. The whole shaft was going banana shaped, flexing out of true by about 6in in each direction. Everything were vibrating — all the shed and the woodwork of the trestles. It was a complete waste of time so I thought we might as well take it down again and put the burner through it. That shaft now constitutes a bracing piece to take some of the strain on the side of the shed.

The only answer now was to have a great big beam about 18in above the proposed line of shafting and a hanging bracket with a bearing on. So it was back to the weaving shed where we had got the shafting from and scavange some of their hanging brackets. We got three of them so we put one in the middle and with a new piece of shafting it all worked perfectly. That section is now about 40ft long and it runs round beautifully.

The next thing were to get another thirty odd foot of it going to where I wanted to build the engine house. I put two more trestles up made out of telegraph poles, with a bit more shafting but this is where I sort of came unstuck and nearly killed myself.

We had got it all up and tried it but it had various wavers and tremors on it, and in fact things were not too good at all with this part. I put it down to one of the particular lengths of shafting myself, because the coupling, instead of being near a bearing, like it should be, were way out in the middle of the piece. This meant there was 9ft 6in of space between the bearings and this big

*Detail of the section in front of the engine shed where the shaft has to be supported from above, by means of a hefty horizontal timber beam.*

cast iron coupling in the middle.

It only needed to be a bit out of balance and it would start bending the shaft when it were revolving. So I decided we would shorten one piece of shafting and get a longer piece the other side so that we could get the coupling nearer to the bearing. I had got two planks from one trestle to the other one and was working up there. I was so mithered that morning as I had had a lot of visitors and ended up making a fatal mistake.

I had slackened off all the nuts and everything and instead of shoving, I pulled. The bloody thing came off at one end and went down and hit the plank with a "bang", broke the plank in half and I went shooting up into the air. Now this particular piece of shafting were about 8ft odd long with a great big 8in diameter cast iron coupling on one end of it. So while I am flying through the air I am thinking to myself 'Where's that going to go?' When you are flying through the air in all directions it suddenly dawns on you that if that thing is going to follow you, one clout from that in the right place and you're dead!

I sort of shot off as far away that way as I could manage, if you can change directions in mid air as you might say, and landed on the floor amongst all the old iron. It were a miracle I did not break my arm or my leg or something, even my neck. I'm sitting there on the ground still rather dazed by it all and still thinking, 'Where's it gone, that piece of shafting?' because I had not seen anything of it since I pulled at it.

What it had done was gone straight up in the air, across the yard and straight down through the top of the shed, and all that you could see were a short piece of it sticking up out of the roof. There was this great big hole in the top of my shed where it had gone through like a rocket, and hit a tin of yellow paint that had exploded all over the shed and all over my taps and dies. In situations like this you look up and think to yourself, 'What am I doing all this for?'. It would be a lot simpler with a little 10HP electric motor. I will never forget that anyway.

I eventually overcame that problem with a longer piece of shafting. On the coupling I had there the nuts were down very deep countersinks. Now I had never had a socket set, all I had were open ended spanners and a few ring keys, so when I put this lot up I could not really get the nuts very tight, but it revolved beautifully.

In trying to get the whole set up right there were a lot of other trials and

*Close up of one of the several trestles made from telegraph poles which carry the shaft from the stationary engine to the workshop.*

tribulations. Like we tried a 60ft length of hose pipe with two glass tubes stuck down the bend and two little hooks, hooked over the shaft to try to get it perfectly level. That did not seem to have much effect. Then we had a piece of piano wire from one end to the other with a big torque bolt at the end so you could go 'doing' and play a tune on it. Then we had plumb bobs hanging from the shaft and that is really how we found out just how far out it were.

It is all right starting levelling from one end but by the time you get to the other end you can be way out. This was as much as ½in out in some places — not sideways, but vertically. You only need to be a bit out with one of them and once the revs get up it will start shaking about. We managed to get it running perfectly and in fact you could not see that it were going round, even when running at full revs. Nothing was shaking, nothing were vibrating. Marvellous.

When Christmas came round my wife bought me this brand new socket set, a box full of all these nice bright shiny silver sockets of all different sizes. On Christmas morning I wanted something to try these out on, so I went out into the garden and looked around, with the socket set in my hand. I remembered that coupling and so I tightened up the countersunk nuts. While I was out there I decided to spend the rest of the morning painting the piece of inserted shaft as it was beginning to turn rusty. This was painted bright red and it all looked very nice now, but the next time I got steam up it were all to buggery again!

The thing is these days they have all these tapered cones and things to tighten a shaft up on so that it is all perfectly circular and symmetrical. It is not just a case of a round hole that is a good fit on the shaft and you bang a key in. That has the effect of pulling it to one side and it has only got to be a few thou out and when you tighten the flanges it puts a set on the shaft about 10ft away. Then, when it is revolving one part starts fighting against another.

What they used to do in the old days was bang the key in and put the whole thing on a great big long lathe, get it running perfectly true and then take a shaving off the end of the flange. Then bolt it together. In the old mills all the shafting would be whistling round and it would look like chromium plate — you could not tell they were revolving other than the spokes in the wheels.

All I could do out in the garden was to just slacken it off half a turn on each nut and it were all right again.

The next thing was to get the stationary engine in position so that it were square with the shaft. We accomplished that with a lot of plumb bobs and by climbing right up to the top of a sycamore tree and having a piece of plastic down spout where the crank were going to be across the engine bed. All this performance were just to put the big stone down for the engine to stand on. I sighted it from the top of this tree, along the shaft to the plastic down spout. You can get things right by doing it that way. Then with lots of plumb bobs hanging down we could see that the flywheel would be vertical and not leaning over to one side.

It was all put into position and the big 6in wide belt was put on while we hoped and prayed we had got everything in the right shop. It worked anyway, it goes round and the belt stays on the big wheel.

I have now got a new 3ft circular saw, to saw up the wood to feed the boiler, to make the engine work, to turn the shaft to work the machinery, including the circular saw that cuts the wood.

The stationary engine itself was found in a mill at Chadderton, near Oldham but it were made in Bolton originally by Bennis Mechanical Stokers Ltd. This engine had driven five mechanical stokers on five Lancashire boilers. The mill was built in about 1904 and that engine were definitely installed right at the beginning and was not a later addition. So it were definitely there since the word "go" and somehow or other it had survived the scrap man. Actually I know why and that were because it were so well hidden behind a door that you could open this door and walk right past it without seeing it. It were like set back in a cubby hole. I went to the mill to mend their chimney one time and that is when I discovered it. I did a deal with the manager and brought it back home here. It were in a terrible state though.

The piston rod were all rusted steel, full of grooves and bumps. The valve rods were the same. The governors had been round that many times that they did not know whether they were coming or going. I had to find some new governors, make two new piston rings, a new piston rod and a new valve rod. As luck would have it the bearings were all right. The gudgeon pin in the crosshead were an old bolt with a split pin through it. That must have come apart at one end because there was evidence that the small end of the connecting rod had been got red hot in a fire and hammered out rather crudely with a sledge hammer. It were all full of these big hammer marks and bangs where somebody had tried to straighten it out. It had probably come adrift while the engine was running and the momentum of the flywheel had bent it on part of the framing of the engine. That was put in the lathe and a shaving taken off it to make it look brand new again.

We tried it first of all on compressed air, instead of steam just to see if it worked, and it did. The compressed air passing through it made it act like a refrigeration plant and the exhaust pipe went white with all these icicles sticking out from it. The vertical boiler was finally connected up and it has been working like that ever since.

Last summer I got a new boiler, a Danks 3-throw 1964 model — getting very modern now. This is in good order and is a lot better steam raiser than the vertical one. I have also got a Wier pump as well so that the water can be pumped for the boiler. Who ever laid that up really looked after it because they greased it up and put a plug in where the exhaust pipe was so that no water could get inside.

The boiler came out of a mill at Oswaldtwistle where I knocked a chimney down one time. This boiler is like new and I did very well to get it for the price, not that many other people would actually want a boiler this big. But in a case like mine I really need it. Well, I suppose I do not really *need* it, but if you are going to persevere with this steam thing you might as well try to get it right.

There is nothing wrong with the old vertical boiler but it were made back in the days when coal was so cheap it did not matter how much you threw on it. When it were made coal was maybe two quid a wagon load and the owner could afford to buy say two loads a week to burn on it. Today a wagon load of coal costs hundreds of pounds. If you over stoke this boiler when it is driving the whole lot of what I have got it will start blowing off. You could easily drive a bigger engine with it, but the amount of fuel it takes is unbelievable.

The big cotton mills back in the days of steam would burn something like 400 tons of coal in a week if they had a 3,000HP steam engine. Steam engines are very uneconomical things in some respects, so that is why they have faded out.

*The vertical boiler with the engine house containing the stationary engine on the right. The belt on the left conveys power from the shaft down to the circular saw.*

## Chapter 4

# More Chimney Pieces

We arrived on the job one Wednesday morning when it was really cold and frosty. We had a wagon load of ironwork for the staging on top of the chimney and everything was absolutely freezing cold. The ironwork was all white and the parts were sticking to each other as you tried to prize them off the wagon. The little knick-knack here though is to put all the pieces on top of the manhole cover on the boiler and it gets hot. When Donald ties it on the rope and sends it up, by the time it gets to me it is still nice and warm and I do not need any gloves. It might be freezing cold but your digits are kept warm.

Soon after we had arrived we looked round the foot of the chimney and saw this ginger tom 16ft up, sat on a packing piece behind the first ladder. We decided we would unload the truck and get all the iron on to the boiler manhole cover, and with a bit of luck the cat might have decided to have come down by then.

When we finished what we were doing it were still up there and it was now sat on the ladder, so I decided I would go up and have a go at getting it down. In after thought I should have gone up it three rungs at a time. I could have caught it easily, because no way would it have risked falling off by trying to go faster than what I could have gone, as its legs were only short. But I didn't. I did not want to scare it, so I crept up the ladder and as I went rung for rung, so it went higher and higher. When I was at the top of the first 16 pin ladder it was at the top of the second ladder, which were about 30ft up. I was beginning to think, 'I aint going to get this bloody cat. Problems'.

I was not over enthusiastic about going on this chimney anyway as it were so freezing cold. I came down and said to Donald,

"I'll go and have a word with works manager about ringing fire brigade and getting one of those teeny-bopper fire engines round".

We did not need one of them great big ones because it was only the height of an ordinary house up. My idea at that time was for the fire engine to swing its ladder round, about 20ft above the cat and then I could go up underneath it. The fireman could then either knock the thing down or grab hold of it. I set off for the office and the manager said,

"Aye, if you ring fire brigade", (the poor fellow is dead now, he had a heart attack not long after all this) "the thing is they'll send us a bill. But if we ring the RSPCA they have an arrangement with the fire service and there is no bill involved".

So he rang through to his girl on the switchboard and told her the tale. He told her to ring up the RSPCA, tell them we have got this cat up our chimney and that we would like them to ring the fire brigade. While we waited me and the manager had a chat about the great days of steam and all that, and how it would compare with modern machinery. We had a cup of tea, as you always do if there is any kind of crisis or disaster. By which time I said I thought we had better go and have a look outside and see what the situation was. He got his overcoat on and as we went out into the mill yard it were like stepping out of a greenhouse into an icebox. The cold seemed to go straight through you. Donald was standing there, looking up at the cat which was now about three quarters of the way up the chimney.

There was another bloke there, with a black overcoat with three pips up. I asked Donald,

"How's the thing got right up there?"

"This fellow here", he said, "he's an RSPCA man and when he got out of his van, looked up at it and said, 'ah, no problem, I'll get it'."

Apparently he had produced this length of conduit piping with a piece of clothes line threaded up it. You need both hands for that so Donald had wondered how he was going to hook it, while climbing a vertical ladder. Anyway he had set off up the ladder with this thing in one hand and as he went up the cat went up further. When he got to about 50ft he had jibbed a bit and come down again, leaving the cat about a hundred odd foot up the chimney.

We were all stood there, all four of us looking up at it when the man from the fire brigade arrived.

"We haven't got a ladder long enough now — you're going to need a helicopter!" It were just a big laugh to him sort of thing.

"Well we've got to have a go at getting it somehow, haven't we?" I said, "Can't just leave the poor thing up there".

As I went up again it went up even further still, until it reached a ledge about 6ft from the top, but this were only about 2in wide. There was not a lot you can do about it 6ft off the top of a chimney belching out smoke.

It was dinner time by now and the RSPCA man came back with a cat box full of cat meat. He said he would tie this to four bits of string, then I could sit on top of the chimney and dangle it down in front of the cat. He thought it would smell the cat meat and jump into the box after it. This box were like a cage! You could see straight through it — the animal weren't going to jump off into space 160ft up, into a box that size. It would never go in there even if it were full of mice.

The cat was on this ledge, the RSPCA man had gone to dinner by this time and the fireman had gone back to base and it were him that 'phoned the television people. When we got back from our dinner the ITV people were already there with their blue Range Rover and they said,

*A telephoto view from the ground of Fred, perched 150ft up on the top of the chimney, December 1983. He vainly attempts to lure the ginger tom from his lofty resting place into the RSPCA's cat box.* *(Courtesy Associated Newspapers)*

"What you going to do about that cat?" and all the usual interview type things.

"All I can do is go up there with this basket and do as I am told".

I climbed back up there yet again and the smoke is coming out profusely and I had a crawl around. It was actually backing round on that 2in ledge, one wrong move and it would have been off. I thought like giving it a kick by this time, but now the eyes and the ears of the world were upon me. No way would it come near me and I chased it round the top of that chimney all

*The scene at the Bolton mill with both Fred and the cat at the top of the chimney. Fred dangles the cat box, but again to no avail.* (Courtesy Associated Newspapers)

afternoon and the smoke were killing me.

I came down at quarter to four, beaten. The thing was still sat there on that ledge over 150ft up then the RSPCA man came out with a classic,

"Tonight, when it gets hungry, it'll come down on its own".

If that was not wishful thinking nothing is. Anyway, we all went home and it rained all night and it weren't too warm neither.

We arrived there at 8.30 the next morning and even by then the BBC television people were already there as well as every newspaper man in the land. Quite a big crowd and all very excited by it all.

"What you going to do this time Fred?" and "Are you going to have another go with the basket?"

All morning I chased it round the top of that chimney, this now being the second day. It just did not want to know. It were very bedraggled by this time and it had changed from orange to nearly dark brown with all the soot and the rain.

I came down and thought 'there's no way I'm going to get it'. No sooner had I got to the bottom of the ladder than the thing climbed right up onto the very top of the chimney. There, I could have caught it because I would be on an equal footing with it. It would not possibly think of jumping down from there, 6ft down the side of the chimney onto the 2in ledge.

Having got down again I was wanting to make a start on the real job in hand, so I decided to start putting the staging up which would be round and underneath the cat. It would then have a platform to come down onto but in the meantime we would just ignore it and leave it up there, after all it had not minded being up there all this time.

We got half the scaffolding up and it would only have taken another two hours but it was dinner time so I started coming down the chimney. At that moment some fellow came rushing out of the crowd below and shouted out,

"We have the Animals in Distress people here . . . "

The manager overheard this fellow and got in a right flutter and said,

"We don't want any heroes here. Nobody up this chimney, bugger off — get out!"

The fellow moved off and I do not even know who he was, I only got a glimpse of him. He definitely was not the chap who eventually did climb up the chimney but probably some sort of spokesman for him.

Being dinner time we went home, the BBC men went to the Farmer's Arms and the ITV men went off somewhere to get some refreshment. One o'clock and we were on our way back. I have been looking at factory chimneys for the past 40 years, since I was a little toddler and I can tell a mile off when somebody is on top of one. It sticks out like a sore thumb. Even at three mile away you can see this little thing on the side. So I remarked to Donald,

"There's somebody up our chimney."

*Doing his best to ignore his feline friend at the very top of the chimney, Fred makes a start on his real work by banging in the first pin for his staging. A few feet above him is the 2 inch wide ledge all round the chimney that the cat had previously occupied for much of its time up there.*

*(Courtesy Associated Newspapers)*

"No there's not." He can't see so well. But I insisted there was.

When we got back to the site, bedlam. There were black marias there, ambulances there and the manager is there screaming away,

"Whose he up there? Where's he come from?" and all of this.

The thing is this chap had actually managed to get hold of the cat and he had it in his hand. But he were stuck. The cat was in one hand and he was holding on to the ladder like grim death with the other.

"Somebody has got to get the fool down", I said looking around. So I climbed up the chimney to see how he was fairing. He were only a young lad and I asked him,

"You all right then cock?" and he replied,

"Yeah, I'm O.K."

The cat were giving him some stick though so I told him,

"Look down between your legs and watch what I'm going to do. You'll be able to leave go with that other hand and change hands with the pussy."

So he looked down and followed my actions of threading my legs through the ladder rungs and being supported just by your legs, and not with your arms. He could then get a better grip of the cat. I asked him if he thought he was able to come down and he hesitantly replied,

"Yeah. I think so." So I told him what we do. I would go down and untie the rope because it having been bit of an emergency over the past couple of days, I had not bothered with that, and it were all tied into every other ladder all the way down the chimney. I explained that when I got to the bottom I would tie this RSPCA man's cat box thing on and I would go back with it. I would open the lid, he would put the cat in, I would shut the lid and he would put the pin in through the shackles to keep it shut.

I climbed back up the chimney and he came down about two ladders from the top. The basket was on the end of the rope with Donald pulling it up. The lad put the pin that holds the lid shut, in the back pocket of his jeans and I opened the lid. He put the cat in, I shut the lid and he put the pin in, all as planned and we lowered it down.

The fumes out of a chimney make your nose run profusely and he turned round and looked at me, and there he was with these two candles of snot coming out of his nose and running right round his mouth and dripping off his chin. He could do nowt about it you see. He'd had too much of this smoke. It has that affect on you I can tell you.

I proceeded to come down and he followed me and we got down on to the floor. The television men were too busy with the ginger tom to notice us and they had all these great big television cameras pointing at it and it looked far more frightened now than when it was prowling around on top of the chimney.

The police promptly grabbed the lad and carted him away and I never saw him again. Though I don't think they charged him with anything. The RSPCA had the cat and not long after it had been brought back to earth they castrated the thing.

But the 'phone calls that Alison had over the two days it were going on were unbelievable. It was actually on the television news the first dinner time of it being up there, and it went out on practically every news and current affairs programme throughout that week. More people saw that one episode, than watched the whole of the TV series that we did.

Anyway, it seemed half the country was ringing up telling us how to get this cat off the chimney. There were some incredible suggestions like,

"What you need is a goldfish bowl with a mouse in it", and the best of the lot, "You want a long plastic pipe and you knock the cat in the end of it with a stick and it'll come out at the bottom".

Number one, where on earth were we going to get a 160ft length of plastic pipe the right diameter and two, when it comes out at the bottom it would be doing 90mph! How do they think them up?

The poor woman who it belonged to lived just over the mill wall and the back gate of her house backed onto the mill yard. With all the commotion and the excitement she were frightened of coming out of the house and she never come near it. Then of course they had to advertise the cat in the Evening News,

"Has anybody lost this cat?"

There were people as far away as Stockport saying it was theirs. It is back home now and believe it or not it used to come in mill yard and sit on a pile of coal and watch us. That job had not really been started when all that happened so we were there for some time after that as we were reducing the height of the chimney. It never went near the chimney again though. It did not want to know about that.

# Death of a Chimney - 1

*Prescot, Merseyside, 25th September 1983. (Photos by Billy Dolce)*

*The 140ft brick chimney at B.I.C.C.'s Prescot Works, as final preparations are made for its demolition by Fred Dibnah using his traditional method of cut and burn.*

The massive size of a factory chimney is illustrated here in this view looking from the ground up. The metal bands strengthening the chimney all the way up are clearly visible.

*Fred shouts a few last minute instructions to his helpers as he removes the final few bricks with a pneumatic drill.*

*A young assistant gives Fred a hand at the square base of the chimney. Material for the fire has already been partly accumulated.*

*Old pallets are placed at the foot of the chimney to add fuel to the fire.*

*The props that now support the whole of one side of the chimney, where the brickwork has been removed a piece at a time. The rubber tyres are used to ensure that the conflagration provides sufficient heat to burn the props away.*

*Paraffin is sprayed onto the timber to make sure it ignites quickly and evenly.*

*Onlookers are kept at a safe distance. Fred's chimney fellings now attract huge audiences with people coming from far and wide to witness the spectacle. In fact it has almost become something of a sporting event and on this occasion bets were being made as to how long it would take from lighting the fire to the chimney falling.*

*The chimney and the surrounding area begin to disappear in clouds of dense black smoke from the burning tyres. The audience, even at this distance, start to feel the heat from the fire and get covered with a sprinkling of hot ashes!*

*The huge fire gets underway, the flames going for a considerable height up inside the chimney.*

About 20 minutes after ignition, and to the exact minute that Fred had predicted, the sound of his hooter is heard along with his now familiar cry of "It's gawing", the chimney begins to waver. The first movement is quite slow, but then . . . .

Within a couple of seconds the 140ft chimney, weighing several hundred tons is crashing to the ground, along the precise path that Fred has planned. The break in the chimney, about half way up is a phenomenon that nearly always occurs.

*The final collapse of the chimney is so quick that it is almost too brief to fully appreciate at the time, so the study of photographs helps even those present at the time, to see just how the chimney has disintigrated.*

*For the second time the area is obliterated by smoke and dust, this time from the crashing brickwork.*

All that remains of a once proud chimney, a pile of broken and smouldering bricks. Fred, being fully aware of the skill and workmanship that originally went into the construction of a chimney such as this, does find such occasions very sad. The only job satisfaction being that it has been brought down without damage to person or property.

A lone works fireman hoses down the last burning embers, while Fred receives the congratulations and applause.

*Fred becomes the centre of attention as newspaper, radio and television people gather round for a few words. He is always happy to oblige with some entertaining observations on the event.*

*Fred has his 'fans' on the works staff who take the opportunity to get his autograph. Several firms who have had old chimneys standing disused for years, have asked Fred to demolish it for them for the publicity value they feel it will give them!*

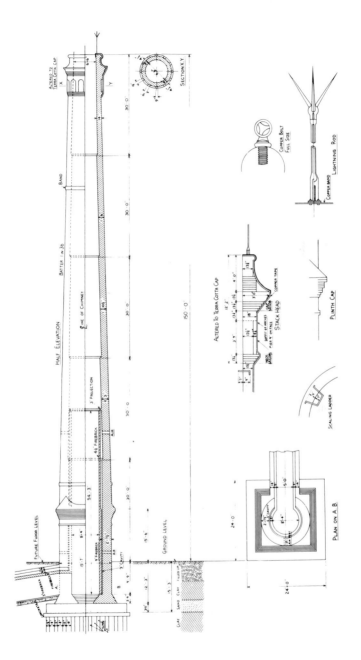

*Cross section of a typical brick built factory chimney.*

# Death of a Chimney - 2

*Platt Ford Mill, Leigh, Manchester. 4th December 1983*

*The octagonal chimney at Leigh a couple of days prior to its demolition. The ornate capping stone is shown here, and note the lightning conductor on the left hand side of the chimney.*

*The beautiful lines of this 200ft chimney are shown to advantage here, with much of the base already cut away. The close proximity of the buildings alongside clearly dictates the direction the fall must take.*

*Fred's constant companion and assistant on his chimney jobs, Donald, closely inspects the props at the base of the chimney.*

This chimney was appropriately, but coincidentally felled on the same day as Fred's first book, 'Steeplejack' was published—4th December, 1983. Left, the fire is well alight and smoke belches out of the top, as it had done for many years previously, but this time it signals its own destruction. Above, it begins to go....

The final second of the fall and the top section of the chimney, complete with capping stone, smashes to the ground, exuding smoke to the very end of its existance.

Half way towards the ground the violent split appears in the chimney. This is a lot longer crack than in the round chimney at Prescot, which in comparison showed a clean break in the middle.

# The Laburnum Mill Saga

We got this job of knocking down this beautiful brick chimney at the Laburnum Mill, Atherton, the whole place being demolished. Alison had this idea of running a raffle with all the proceeds going to cancer research. The person who held the winning ticket which would be drawn on the morning, would help me light the fire at the bottom of the chimney, then retire to a safe distance. This is something people are always wanting to do and must be something to do with a destructive element in people's nature or something. It all went wonderfully well and we had already raised over £1,200 selling the tickets and there had been an enormous amount of local publicity for the event. There was even a chap coming up from Radio Severn Sound in Gloucester to record the event for radio — we had had several of our chimney fellings on television, but not on the wireless before!

It was the night before the Sunday morning we were due to bring the chimney down and we were just about to go up to bed. Then the telephone rang and the woman on the other end says,

"Somebody has set fire to your chimney stack."

I immediately thought, 'this has got to be a hoax. It's quarter to one in the morning.' I thought 'definitely not on'. I asked her if she had a boss I could speak to and as soon as she laid the receiver down I could hear typewriters and other telephone bells in the background. I realised that it was either a fire station or a cop shop. This other fellow comes on and he says,

"It has been burning from about 12.30."

I asked him,

"Is it still up do you know?"

"Yeah, just about. The fire brigade are there now. Can you get over there now?"

"I'm on my way", I said. It were about five miles from here and I drove at top speed as there was nothing on the road at that time in the morning. When I got there I could just about see that it were still up in the blackness, the fire being almost out. The fire engine was parked outside the gate of the place, at least 200 yards from where the chimney was.

"Can't you get over there and put fire out?" I asked them.

"We're not going near it", they said, "there's a big bottle of propane stood right in the middle of the fire".

I was amazed at this and said, "That propane has been there since the fire was lit at 12.30. It's 1.15 in the morning now. If it ain't blown up yet, it ain't going to blow up".

By now it was quite black over by the chimney and there was not even any glare from the roadway. I borrowed a torch from a policeman and went over and had a quick look round it. When I got nearer the front I could see that of the twenty five 8in diameter props there were only three left on each side and they had suffered from the heat. This bottle of propane were not pink any more, it were now a dirty grey colour. There was nothing issueing from the end of it and it were quite dead now.

It was evident that what the vandals had done was they had opened the top on the propane and let it blow out and thrown a lighted match into it. That is all the combustible stuff there was because during the course of the Saturday afternoon we had shifted all the small timber so that nobody would get the idea to light the fire.

It was still up but then I wondered round the back and shone the torch on the horizontal crack. There was now no way we could keep it up. If it had been stable I could have stuck another prop under the middle and that would have saved it. It would have held it up until we wanted it to go, but it were too late. The crack was about 3/16in broad at the back. When we had left it on Saturday afternoon it were only 1/64th of an inch and about 8ft round the circumference. Now it was practically half way round and at the worst part. 3/16th of an inch on a chimney that height means it is about 18in out of plumb and it was ready for going. Within another ten minutes it were creaking and groaning. I advised the fire brigade, who had got brave enough by then to come near it, and the cops that had advanced cautiously down the road, that we had better all back up because it was going to go. We had only gone about 100 yards and "whoof", down it went at quarter to two on the Sunday morning.

It were rather sad really because we caused a traffic jam on the Sunday morning and there was no chimney there. People had travelled from as far away as Sheffield to watch this chimney come down. They never got anybody who was responsible for it though they reckoned they had their mits on some who had been split on. Even then they said,

"What can we charge them with if we do get them?".

That whole place seemed to have a kind of jinx on it. At the other corner of the mill there was this great big brick tower, 170ft high. I had never done one of them before but we got it already for bringing down. Even during the preparations we had had a slight tremor.

The walls are very thin on these things for the weight they are holding up. We started cutting at this wall in the front, this being the way we wanted it to fall. I started at one end and another chap started at the other and we advanced along the wall towards each other. The wall was approximately 2ft thick and we had six or seven props each side of the cut as we progressed towards the centre. When we got to about a 2ft square in the middle this pillar of bricks began creaking and moaning and vertical cracks began appearing in it. Bits kept blowing off it and for three quarters of an hour we just stood there and watched this pillar go from 2ft square, as though invisible beavers were going at it, down to a 4½in square.

All the weight had been on this centre pillar with next to nothing on the props but now the wedges had been squeezed to oblivion. There was now just this little piece left in the middle and I hit it with a hammer, "boom" and it simply fell out. There were these two beautiful pyramids of crushed brick which had appeared before our eyes.

Our troubles were not over by any means. On the Sunday morning we lit the fire and there was a big crowd there as usual, including the people from German television who were making a film about us. A beautiful horizontal crack appeared across the back of the tower — marvellous, everything were in our favour, apart from the wind which were blowing in the wrong direction.

There were one very weak corner on the tower and one very strong one and the wind was blowing towards the weaker one. The fire was therefore being blown towards the corner so that meant the timber there would burn away first and the timber on the strong side would stay intact the longest.

It kept burning and it actually started to go, the whole lot of it, but then about 200 tons of masonry and bricks fell off the front. It then just sort of sat back on itself and it stood there, unmoved. There was this ginormous hole in the front of it, it were terrible when you looked at it. The fire was still raging away underneath the brickwork that had fallen down. We had to get the fire brigade in to put the fire out. I then had to nibble out a bit more at the bottom, no more than another foot or so and it was away again and down it come. It was all very exciting and spectacular. We now know what to do if we get another one.

## Chapter 5

# The German Connection

We had got used to the BBC people following us around sort of thing. They came with me to work, more or less moved in at home and even went on holiday with us when we went to Blackpool. But it came as something of a surprise when we had a letter from a German television company, ARD saying that they wanted to do the same.

They only wanted to do the one programme at that time so we took them to see the tower at Laburnum Mill coming down and they came with us to a traction engine rally. We got on with them quite well really and they sent us a copy of the film later and that did make us laugh. It was all in German of course, except that every now and again the commentary broke into a few English words — just the swear words, they seem to think I use occasionally.

We were only part of the final programme which were called "Rund um Big Ben". From what we could make out it appeared to be about English eccentrics, so I don't quite know what we were doing on that.

They have now been wanting me to go over to Berlin so I do not know what will come of that, but it would not be my first experiences of that country. I started to tell that story before, in the previous book, so perhaps it is appropriate to recount that now.

When I was called up into the army I pleaded with them to give me an outside job, like building or engineering. But what I got was the cookhouse and when I heard that, I nearly deserted. I just could not imagine myself being a cook for two years. Fred Dibnah, Cook. It did not bear thinking about. Out of the 40 of us the army took on as cooks, there was not a single man who had previously had anything to do with cooking.

One fellow had worked a steam hammer in Sheffield. There were a lot of bricklayers from all round the place and a couple of joiners like myself. The nearest to the food business was a chap who had a knackers yard. But the army had made up its mind and like it or not we were to be cooks and what better place to teach us than Catterick Camp.

It were a dreadful place and our cooking lessons no better. I could not get away from the place fast enough on my one weekend off. I practically walked from Catterick to Bolton which is a long way. As I neared home, I spotted some steeplejacks mending two big chimneys which were side by side. It meant a long detour on top of all my other walking, but it did my heart good to see them chimneys with red ladders up them.

When you come to think about things, it is a funny world. That weekend leave were the low spot of my life. No sooner had I reached home at Bolton than I was on my way back to Catterick. I saw those two chimneys again with the steeplejacks repairing them and could have wept. Years later, when I were a real steeplejack myself, I brought them both down for BBC television.

However, back to the army days. After finishing our cookery course came the great moment of volunteering for where you would like to go. I put my name down for Germany because some old soldier had told me things were more lax abroad. Anyway, I was loaded on a train, a boat and then another train. Eventually I ended up in a place like Belsen. It were a huge camp for tank regiments which someone said Adolf had built up for his many men.

Of course my place was in the wretched cookhouse. They told me to forget all about my training and concentrate on great dustbins full of spuds which were full of black eyes. So the potatoe peeling section was where I spent most of my time. It were full of cockroaches as was the canteen. They used to creep through the edges of doors and liked to rest in the curtains next to the dinner tables.

I have knocked around a bit myself but I found those army boys a rough lot. If there was a bit of barney and one did not like the other, he would flick a curtain as he passed the end of the table. This sprayed cockroaches not only into his enemy's dinner, but covered everyone else's food. You knew then that you had joined the professionals.

One slightly good thing about the cookhouse lark was that if you did breakfast and dinner, the rest of the day was your own. I did not mind starting at four in the morning because it gave me the afternoon to myself. I took to getting as far away from the camp as possible. The area was covered with big pine forests — Christmas tree jobs — where you could walk for hours and not see a soul. I used to tramp along thinking of my home and parents, Bolton and all them chimneys. Would I ever become a steeplejack and get red ladders up them? Would the name Fred Dibnah ever become as well known as my boyhood hero John Faulkner? These were dreary walks and thoughts.

The 14/20th Hussars as they were called had only just arrived in Germany. They were supposed to be a cavalry regiment but had tanks instead of horses. However the officers were like sons of wealthy farmer types and they had arranged to bring their own horses with them.

One day I were proceeding along a forest track and came upon a barn with a big hole in its roof. In Germany all the roofs have a very steep pitch and gazing up at it was one of the officers from the tank regiment. Apart from his army boots he did not appear too smart. In fact he looked as if he were back home beside his hen pen. I bid him good-day sort of thing then he began talking about the steep roof and the difficulty of mending the hole.

When I told him that I were a joiner, a steeplejack and a lot more else he wanted to hear, there was delight all round. Soon we were joined by the major, who acted as God back at the camp. He were six foot six inches tall and had a moustache almost a yard wide.

"Do you really mean", he said, enthusiastically, "you can mend that roof for us? You see our horses are on their way and we must have this barn to keep them dry and comfortable".

"Of course, sir", I assured him. "I can mend your roof if you can get me out of the cookhouse".

"Say no more", he said.

When I got back to camp, the monster who were in charge of the cookhouse were all smarmy like. He told me how pleased he was that I would be helping the major and other officers to house their horses. He had always thought me a good lad and all that.

I did the roof job as slowly as I could. When I was nearly finished the captain asked if I could lay bricks. You see, they now wanted stalls built for their horses. Well, I again assured him of my capabilities and in due course loads of bricks kept arriving. It suited me because the months were ticking by.

When I had finished the stalls, I spun out the job doing anything to please these horse mad officers. My masterpiece were a traditional British weathercock which I fitted to the top of that German roof. It were made of battered up army trays, spoons and things, all welded together.

There was nothing else I could make or fix around there and it looked as if I had to go back to the cookhouse when, lo and behold, I seemed to have another stroke of luck. Twelve British fox hounds were arriving to join the horses, so they needed kennels and someone willing to look after them.

I have never been particularly close to animals and some of those hounds were ferocious. Heaven help the poor foxes who crossed their path. Worse still, the hounds were always hungry and even the local German butchers had problems supplying them with pigs feet and intestines, all in messy dustbins.

Instead of feeding men I became a dog cook — stuffing this offal and hard army biscuits into what they called a field boiler. That were bad enough, but afterwards came the dangerous job of going into the compound full of great big hounds. They seemed to think I were part of their feed and would even fight over me.

It were a bit of a problem for me, but one day I ran into this sort of hound expert. He said,

"You know what you want to do. You want to take a pitch fork in there with you, then lay the handle across the first hound that comes near you. If you hit him hard enough, he will see sense and so will the others".

So I did what the hound expert told me. The first beast got the handle across his backside, which made him turn on me all snot and teeth. I had to hold the pitch fork in front of me while thinking 'one false move and you're dead'. Meaning myself of course. But the expert were right. That hound backed off and the others seemed to get the message. The next problem was when these officer farmers took their horses and hounds out hunting. Two or three dogs went astray each time. This meant yours truly had to go round the forest like a proper Charlie hooting on a hunting horn. We were sort of helped by the German gamekeepers. They used to return our dogs on the backs of motorbikes — after shooting them dead.

I did not worry too much about the number of hounds getting less. So was my time in the army. My two years national service was nearly up when a Mr. Profumo had a brilliant solution for the lack of recruits. I had just one more fortnight to go when he announced that we national servicemen would have to stay on for another six months.

There were about 40 national servicemen in that particular camp and some of them had children they had never seen. You can imagine how they felt while I felt the same about my chimneys. After all we were only playing at being soldiers. No one minded the real thing but this meant wasting more of our lives. There was a hard core of resentment about the camp and the army as a whole.

Little was printed in the local papers because the War Office acted very craftily. They invited the press to send over reporters who would ask national servicemen their opinions on the extra six months. Of course it were all done in front of officers with NCO's right behind you breathing down your neck. Like all the other lads, I were as sick as a pig about it but I said,

"Well, if it's got to be done, we're good lads and we'll do it".

"Good lad", said the officers, while the NCO's seemed to breathe a little easier.

After this mockery of interviewing, the reporters got a big booze up in the officers mess so that there was very little printed about how hard we were hit and suffered. No sooner had the reporters departed than the national servicemen were dispersed all over Germany. Only two or three were put in each camp full of regulars and treated a bit like lepers.

I was sent to another camp about 20 miles away for the final six months and worked in its cookhouse. During my time off I could lie on a campbed thinking of what to do when I finally left the army. Should I return to being a joiner or should I have a real go at steeplejacking? I were then 25. I had done my bit for my country bit it had not done very much for me.

While I were lying there mithering my brain, I invented a lot of things that would improve steeplejacking. For example I thought up Dibnah's famous flying buckets, which I described in the previous book, whilst lying on that campbed in Germany.

So it looked as if I were going back to my first love — chimneys and steeplejacking. To tell the truth, there was never any doubt about it.

## Chapter 6

# The Famous Fred

My chimney performances have attracted quite a following. Pensioners, children, strangers who describe themselves as chimney enthusiasts and citizens who in former times might have attended public hangings. They all now turn up in the hope of seeing a disaster.

I can assemble a crowd which would be envied by Bolton Wanderers. I play up to them a bit, act as if the damn thing is going to come down on top of them, or on a building. But no way is it going to do that. I have taken out half the bricks on one side and that is the way it is going to fall.

"Well done, Fred", they say when the dust settles.

"Magic"

I have never had a disaster nor is there any reason why I should as it is all carefully worked out beforehand so that I know exactly where it is going to fall. But try telling that to the insurance companies. On one job I got paid £200 for bringing the chimney down. The insurance company got paid £700 in case I cocked it up. I had done all the hard work a week before and they never even set foot near the place.

One time I did not have two ha'pennies to rub together and nobody was interested in me. Then I risked life and limb. And still nobody cared. Then the television people showed what I did and now I get letters from all over the place.

The things people want to give me are incredible. One fellow offered me five lathes. Someone else wrote,

"Dear Mr. Dibhah, We are very concerned that your matches keep falling out of your pocket when you are on top of a chimney. Enclosed is a sample of our product — wind proof and weather proof matches".

I once had a 90 year old lady come round to my place on a Sunday morning, delivered by her son. She said her husband who had died was into steam engines. Watching us on television with our steam engine had made her happy.

"It's the first she's smiled", said the son "since father died".

On another occasion we were out with the steam roller and we stopped outside this house and the lady came out and she was crying. She came up to me and said,

"Oh you gave me a fright. I thought you were my Harry come back from the dead".

Apparently her late husband used to drive a green steam roller like ours and he used to park it outside the house where I had stopped, and at a distance I didn't look too unlike him. She invited me in for a cup of tea and showed me some photographs of him and a couple of little instruction books for steam roller drivers he had left her.

Unfortunately I did not have my glasses with me at the time so I said the next time I was passing I would call in and have a read of them. A few months later I was passing her door and she saw me and beckoned me in again. We had another cup of tea and she said I could keep the books as she knew I would be interested in them, but felt that when she passed on her family would not value them but just chuck them out. They do make fascinating reading so I have included a few extracts from one of them as follows:

**"Instructions to Steam Road Roller Drivers."**
The Lancashire Road Roller Co. Altrincham.

It is interesting to note that before Harry Hulse had the book the previous steam roller drivers to have used it had been a P. Abram, A. Moolley and a P. Johnson. In the back of the book was Harry's wages slip well tattered and torn from 1933 and I am sure his widow will not mind if I quote this now as well.

*Instructions to Steam Road Roller Drivers*

No. *1 by*

Driver

H. Hulse.

The following Instructions must be read, and remembered, by all drivers, as in every instance failure to comply with them renders either the driver or ourselves liable to a heavy fine.

The Lancashire Road Roller Co.

Altrincham.

This Book is the Property of THE LANCASHIRE ROAD ROLLER Co., and Drivers must RETURN IT when leaving their employ.

*The Lancashire Road Roller Co., Altrincham.*

3

QUESTIONS BY STRANGERS.

On no account must drivers discuss with strangers particulars of their work, where they are travelling to, or where they have been working, or any information concerning their present, past or future work. Firms competing with us have endeavoured to obtain such information to take our work from us, and we expect our drivers to assist us in every way they can to retain this work.

ROADSIDE WASTE LAND.

Steam Road Rollers, Vans, and Watercarts, must not be parked on Roadside Waste Land, unless permission to do so has been obtained from the local Surveyor.

Although this regulation applies in all parts of the Country, it is particularly enforced in Derbyshire.

## TRAVELLING.

### 1. FLAGMAN.

When a Roller is travelling or standing on any highway, a second man (or flagman) is required to be constantly with the roller.

If for any purpose the flagman leaves the roller, the driver must at once cease travelling and obtain assistance from the police. One person, however, must remain with the roller when on a highway, so long as the fire is alight, or whilst the roller has sufficient steam to move itself.

### 2. WATER.

Rollers travelling through Manchester and Salford are allowed to take water from the Mains (for boiler purposes only).

Rollers travelling through Bury are not to take water from a Public Fountain, trough, well or receptacle for water, except with license of the Corporation

**WATER** *continued.*

In all other places permission must first be obtained before taking water from mains, ponds, streams, reservoirs, etc., from the owners, except from recognised watering places.

If a driver finds his roller urgently requiring water, application should be made to any policeman for assistance.

### 3.—SMOKE.

It is an offence punishable by a fine of £5 for a Steam Roller passing through a town or village to make smoke.

As far as possible, when travelling through populous districts, coke should be used. Otherwise, the roller must be fired outside the towns or villages so as to avoid making smoke during the passage through.

### 4. REGISTRATION NUMBER PLATES.

Every Steam Road Roller is now compelled by law to carry two registration Number Plates, one at the front and one at the back of the roller, in such a position, and kept clean and uncovered so that they are easily visible, and can be easily read by the police.

When hauling a water cart, sleeping van, or any other vehicle, the last vehicle must show on the rear a Registration Plate, bearing the same number as that on the engine.

The penalty on the driver for not having the Registration Plates, on the roller or trailing vehicle, or for having them covered or obscured, is £20 for the first offence, and £50 for each subsequent offence.

A Steam Roller is also required to carry the Paper License as carried by motor cars and traction engines, and this must be fixed to the side plate of the engine, next to the Name Plate, and kept clean

### 5. NAME AND WEIGHT PLATES.

The Roller must have fixed to it in a conspicuous position :

(a) The Name-plate giving our name and address.

(b) A Plate giving the nominal weight of the engine.

The Van and Water-Cart must each have fixed to it in a conspicuous position :

A Plate giving with weight of the vehicle.

### 6.—OBSTRUCTION.

It is an offence to purposely obstruct the passage of any highway, and drivers must give as much space as possible to other traffic.

Special attention is drawn to page 18 giving times a roller may stand, and also about passing through narrow streets.

### 7.—SPEED.

The speed of a roller must not exceed four miles per hour in the country, and two miles per hour in cities, towns, and villages.

78

## 8.— STEAM.

The Cylinder Taps of a roller must not be opened within sight of any person with a horse upon the road.

Steam must not be allowed to attain a pressure exceeding the limit fixed by the safety valve, and so blow off while the roller is on the road.

## 9.— STOPPING WHEN REQUIRED.

The Roller must be instantly stopped on the flagman or any person with a horse or carriage putting up his hand as a signal to stop.

## 10.— LIGHTS.

Every Roller on any highway or road, to which the public have access, must, between sunset and sunrise, carry lamps as follows :

(a) At the front, two lamps displaying to the front a white light.

(b) At the rear, a red lamp displaying to the rear a red light.

(c) If hauling a van, water cart, or other trailing vehicle, a red light must be displayed to the rear on the back of the last vehicle.

## LIGHTS *continued.*

Note These regulations apply whether the engine is stationary or moving, and drivers are warned that when stopped for the night the roller must not be left on any highway or road to which the public have access.

Warning—We do not supply lamps with any of our rollers, as on no account are drivers to travel or leave their engines on the road at night.

## 11.— BRIDGES.

Care must be taken to observe the regulations regarding the maximum loads on bridges contained in notices posted on or near bridges.

Rollers must not pass other rollers or locomotives on any bridges ; one engine must stop to let the other pass over.

A driver shall not allow his engine to remain stationary on any bridge, arch or culvert, or any approach thereto. If thereby a driver is unable to pick up water, the water cart can be taken on to the bridge and filled up, and the water taken from the cart into the engine.

## 12.—HAULING VEHICLES.

A Steam Road Roller is not allowed to haul any vehicles other than its van and water cart, as it is not registered as a haulage locomotive. The penalty for not complying with this is £20.

## 13. ACCIDENTS.

In all cases of accidents to the roller or driver, in which another vehicle or person is concerned, the driver of the roller must :—

1. Make a full report in writing to the Office at Broadheath the same day, giving full particulars of the accident.

2. Refuse to make any statements to the police or any other persons. All enquiries of this nature must be referred to the Office. This is most important and must be strictly observed by the driver.

## 14.—GENERAL.

1.—A driver, if stopped by the police for an alleged offence, or in the case of an accident of any sort, should obtain the name and address of an independent witness.

## GENERAL *continued.*

A driver is only required to give his name and address, and answer other direct questions if asked to do so. **He must not make any general statements.**

## 15. LIVING VANS.

Drivers are requested to note that Living Vans are provided for their own accommodation only.

Under special circumstances no objection will be taken to a man's wife occupying the van for short periods.

**Under no circumstances** are children to be housed, and any contravention of this instruction will mean dismissal of the driver.

79

# WAGES FOR DRIVERS

Driver: H. Hulse 1933
Your wages for week (Hours to be according to the Council by whom the Roller is employed)
£2 :1 : -.

**BONUS:** Rolling and first day's travelling, 1/6 (one shilling and sixpence) per day (half bonus only when Saturdays count as half days).

**TRAVELLING ALLOWANCE** at Flat Rate of 1/6 per week, except when Roller is standing owing to completion of job, or when Driver is in Works, when no travelling allowance is paid.

**LODGING ALLOWANCE.** When Van is not provided 5/- per week.
Tandem and Footpath Roller Drivers 10/- per week.

**WASHING OUT.** When the steam roller boiler is washed out, out of working hours, 4/- will be paid for carrying out this job. (The Rolling bonus will not be paid for washing out).

**HOLIDAYS.** On the undermentioned Public Holidays only, all drivers will receive half wages where the Council or Contractor stop work:-

| | |
|---|---|
| New Year's Day | Whit Monday |
| Good Friday | August Bank Holiday |
| Saturday after Good | |
| Friday | Christmas Day |
| Easter Monday | Boxing Day |

**STANDING.** For the first 3 days only at Standing Rate. Drivers must advise Office on the second day for further instructions. No other standing time will be paid for.

**Lancashire Road Roller Co.**

Broadheath, Altrincham.                    Tel. 2645 Altrincham
February 1933

80

Since the television series was shown, in addition to being given things we also get asked to appear as what they call the 'guests of honour' and to officially open things. At one time this was just a local thing like a Dr. Barnardo's shop but now it can be all over the country.

We get invited to lots of traction engine rallies and if we went to all those we would never get anything else done so we often have to refuse. Even those we do go to are often too far away to take our own engine as I do not agree with the idea of sticking it on the back of a lorry. If we cannot drive it all the way there under its own steam then it has to stay at home. Last year they invited us down to Cornwall so I went with Alison and the three girls to the West of England Steam Society's Rally at Silverwell Farm near St. Agnes, and we officially opened that.

This year it was the Holton Working Steam Rally near Oxford. While we were there for the weekend they loaned us a very rare Richard Hornsby engine, "Maggie" which dates back to 1886.

People are often requesting us to visit them and then they say, "Don't forget to bring your steam roller with you Fred".

These people do not realise what is involved in firing up a 12 ton steam roller and driving it out on the street and then putting it back in the shed again and cleaning it all out again. They seem to think it is like turning a car ignition on. We have even had a request from a bookshop where we were going for a signing session to take the roller along. I thought we would have to sell a lot of books just to pay for the coal so we went in the Land Rover instead. When we got there the shop was in a pedestrian precinct and we could not even park that anywhere nearby.

Other people turn up and knock on the door and expect you to get the engine out of the shed so that they can take a photograph of a relative standing by it. Then the other day some more television people came on the 'phone and said that they wanted the engine in steam standing outside the shed so that they could film a few details of it. I said, right that will cost you so much.

"We can't afford that!" they said and offered me about half. Eventually we agreed on a figure and they came round. When they arrived there was about six posh limousines and each chap wore a fancy suit and had a dolly bird assistant with him. Each of them were fancily dressed in pink silk overalls. All in all they did not give the impression of being too hard up.

I got the engine out of the shed for them and they started filming various parts of it for some technical programme, something educational anyway. Then one of them shouts across,

"Hey Fred. Can you come over and tell us what this bit does?"

So I went over and started explaining, and the cameras kept rolling. Next thing, we saw Alison rushing out from the house and she says to them all,

"We agreed a price to film the engine. Now you are interviewing Fred for

*A mass line up of steam engines at the 10th Working Steam Rally, Windmill Field, Holton, near Oxford on 9th June. Nearest the camera is a local engine Tasker 1666. Fred and his family attended this event as the special guests.*

*Alison often takes a turn at the wheel when the Dibnah's own engine goes on the road and she is seen here on a Richard Hornsby built engine at the Holton event in 1984.*

*Fred at the controls of Richard Hornsby 6557 of 1886 (registration FL 2598) at Holton. This engine, which was at his disposal for the weekend, is particularly noteworthy as it is now the only known surviving traction engine of this builder. (The firm later became Ruston & Hornsby Ltd upon merging with Ruston, Proctor & Co. Ltd on 11th September 1918). Livery is unlined red with polished brass boiler bands.*

television. That's an extra twenty five quid". We got it too!

I get asked to go to some unusual places but one of the more interesting ones was to open a new telephone exchange for British Telecom at Swindon. The general manager of the telephone area happens to be a steam enthusiast himself and his son regularly drives a friends traction engine. They arranged all the transport and everything and I arrived in Swindon where a beautiful Burrell traction engine were waiting for me. It were a beautiful day and we chuffed all round the town with a little trailer on the back which carried a board reading "From the Steam Age to the Electronic Age with Fred Dibnah". This trailer carried extra fuel and some people from a local radio station who broadcast everything we said.

We had a fine old time clattering and whistling round this town, once so famous for building steam locomotives for the Great Western Railway. But then the crunch came. We arrived at the exchange and the yard they wanted me to park the engine in only had a narrow gateway. No way would I get in there as there was a brick wall right across the end of it. As we came round the corner and approached it I asked, "What do you want me to do now?"

*Fred clatters through the streets of Swindon, Wilts on Ted Haggard's 1910 Burrell, number 2426 (registration BL 4843). They are en route to open the new electronic telephone exchange at Regents Close on 5th July 1984.*

*The engine rumbles on with its tender behind, containing coal as well as local radio transmission equipment — note the tall mast mounted on the engine footplate.*

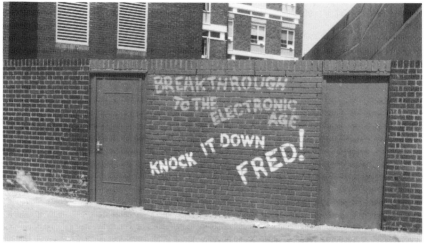

*(Courtesy British Telecom)*

"Just keep going Fred" they said "This engine should get through there without too much trouble — open her up".

'They know what they're doing' I thought so with a blast on the whistle we opened the regulator and went straight at the wall, "Crash". We were right through it with bricks flying in all directions and a cheer from the people that had gathered round.

*(Courtesy British Telecom)*

*Crash! — the engine breaks through the 'brick' wall as it enters the telephone exchange yard. Fred is assisted on the footplate by Colin Hatch, whilst the radio man holds the microphone close to him to make sure he does not miss any choice comments from Fred.*

I was a bit disappointed really when I realised the wall was a dummy made of polystyrene bricks, but it was very convincing. I then had to make a speech before cutting the tape which formally opened the premises. This was about three floors up, on the outside, so I was whisked up in a hydraulic platform. Then we had a wander around inside and saw all the electronic wizadry. It was all a very far cry from the Victorian mills of my home town of Bolton.

I also get asked to appear on various television programmes these days including an early morning talk on TV AM and we had the girls on "Whose Baby?" a few months back. Then another time I was the judge on "The Great Egg Race" as the three teams were asked to make a steam engine from the various bits and bobs they were given. They asked me first of all how I would do it. I looked around the things and had to admit it were rather difficult. How do you make a steam engine when lots of the parts are plastic? Anyway, about six hours later the teams had concocted various devices which were powered by compressed air rather than steam. In the meantime they got me to explain how a real steam engine worked by using an Isle of Man loco that had had one side cut off.

My most surprising television appearance though was not me in person at all. It was Mike Yarwood impersonating me on his Christmas show last year. Now you can't get much more famous than that can you?

*Time for a breather and a drink. Fred and Andrew Hurley, General Manager, British Telecom, Gloucester Telephone Area, pose for photographs in front of the engine after its 'break through'.*

*Chapter 7*

# The Rise and Fall of King Steam

People often ask me why it is that there were so many chimneys in this part of the country and why it is I am now bringing them down. To understand this you need to know something about the history of the weaving and spinning industry in Britain.

Humanity in general has been weaving and spinning material for centuries, but for a long time using very crude methods. For some unknown reason Samuel Crompton, Richard Arkwright and other chaps like them with inventive type brains, all lived in the area. Crompton who lived in Bolton, invented the spinning mule in 1780.

The business men of the time started constructing these things and of course began to make the money. It was such a fine piece of tackle that these business men soon realised that if you had say 20 of these inside a shed, and a lot of little lads working them, you could make yourself a fortune. The other aspect was that there were plenty of coal about nearby which were an advantage, because in them days transport were a bit awkward. That is basically how the cotton era started.

It all grew and grew and developed; steam engine builders, boiler making companies all got bigger and bigger. About a hundred years ago the mills in Lancashire employed about half a million people, producing something like 10,000 tons of cotton every week.

By the 1900's the spinning mills were at their finest. Most average mills had four boilers and a 2-3,000HP steam engine that turned it all round. Some concerns were even bigger and they had three or four mills with two or three engines and two boiler houses and two great big chimneys.

A lot of the owners were so far seeing that they would purchase the field next to their site with the idea of extending and building another mill. Many of the mills owned a huge amount of land but now it is all being sold off for housing estates and what have you.

They seemed to have this idea at the time that they were going to last forever. You only need to look at the architecture of the buildings to see that. Some people refer to them as the dark satanic mills, but you want to look at some of the rubbish they are building today! They have recently cleaned up some of the mills and they really are beautiful, with all their fancy twiddly bits.

Some of the mills were completely self contained and they even had their own bore holes for water, their own blacksmiths shop, joiners shop and mechanics shop. The only thing that they bought from outside was the coal. By the 1920's many of the engineers began to realise that the great age of steam was nearly finished and the electric motor started to take over fast.

Going back to when I was about 14 years old, as well as steam engine builders round here there were spinning and weaving machinery builders that were not too far away — Mather & Platt, Dobson & Barlow and lots of others and they were world beaters in their field. Their backroom boys were still improving and making better machinery. By then the mills that were still privately owned were now in the hands of the third generation of owners. These people were born into apathy with silver ware on the table in great big mansions. They took it in turn all round the town to be mayor and all that sort of thing. One half of the family built steam engines and boilers while the other half owned the spinning mills, so they did not want to modernise the places. One family owned about seven mills (I've got my ladders up one of those at the moment) and that was probably the biggest complex in Europe. Originally they had seven steam engines but eventually they scrapped these little ones and got in bigger ones that would power a couple of mills. They knocked all the little chimneys down and built two great big ones.

I have got one of the iron plates off one of their engines, built in 1888 — "John Musgrave & Sons, Ltd, Engineers, Bolton". That engine worked right up to 1960 driving a big mill. They simply never modernised. Dobsons kept building the machinery that would keep them in the forefront of the world, but the mills did not want to buy it. They still had their own ancient mules rattling up and down on knotty pine floorboards.

The first blackmen I ever saw were the sons of rich merchants from afar, and they were all at the technical college across the road. There they were learning all about the cotton trade and the machinery. Then they went back home, daddy bought them Dobsons latest machinery and got them going. Eventually they thought their machinery was getting out of date so they went and bought Swiss and German machines which were improvements on the others.

Now to compete with the foreign competition Courtaulds have had to buy new machinery. Mather & Platt have gone, Dobsons have gone so they have had to go and buy the Swiss and German machines. Nearly every working mill has foreign machinery in them, apart from the carding engines. Some of these still have "Mather & Platt 1915" on the castings on the ends but they are totally different inside to what they were in 1915. They have got new guts inside the old frames where the mills have had a go at modernising themselves now.

All this foreign machinery in the mills that are left is now working day and night and never stops. There is only about ⅛th the people that were working there and they are more like robot places.

*Queen St. Mill, Harle Syke, Burnley, Lancashire—the last working steam powered textile mill in Britain.* *(Courtesy Queen St. Mill)*

One mill after another shut that would not modernise. There was once about 200 in Bolton, now you can count them on one hand. There is one privately owned, or at least owned by a limited company and still trading under its own name, but the other three are all owned by the giant Courtaulds. Tootals, famous for their shirts and ties once had a whole empire here, with about five chimneys on.

I remember Princess Anne coming on one occasion and saying something like, "They've modernised now. The future is all bright and rosy". That place is flat now, it's a cinder patch.

Once they said "England's bread hangs on Lancashire's thread". There was more money made in these parts than any other part of England, except perhaps the steel towns like Sheffield. But as for the concentration of boilers and chimneys there could not have been any place in the world that had more. You need to think of the sheer power in horsepower contained within the town at the height of the milling industry. There were dozens of 3,000HP engines. Some of the buildings are still there but with something like a washing machine motor in them.

They were very proud of their engine rooms and they were kept spotlessly clean. The door was always at the top of a flight of steps and it was always very well joinered. It was more like cabinet made than joinered and they always had brass key hole plates and knobs on. I now have a pair of these brass door knobs on the door of our living van.

No one was allowed into the engine room, it was taboo. Even the manager of the works himself was scared of going into that forbidden place as the bloke who ran the engine were like a little god. Those engines were really magnificent machines, even if in reality they were very uneconomical.

The last two steam engines in Bolton must have been among the very last of the real big ones to actually be used commercially. They were at Warmsleys Forge and these two were 30ft tall, vertical single-cylinder rolling mill engines and ran right up until April 1983. When the firm started in the 1860's they had two steam hammers and three rolling mill engines. When the first of these were removed it was not scrapped but went to the Ironbridge Gorge Museum in Shropshire, complete with one of the steam hammers. That well known museum has acquired a lot of its tackle from the Bolton area.

There are quite a lot of stationary steam engines preserved in museums, but to me these are merely mounted and stuffed, they are not what I call working any more. Some of them have their 'steam days' when they light the fire and they make the flywheel go round, and even if they have a few ropes on they are only running a pulley up in the rope race. The engine only goes "chuff, chuff" as there is no shafting upstairs for it to really get to grips with.

When the engines worked at full pitch they had a 100 tons of weight on them and you could feel the vibrations, very different to when it is out to

*The 500HP tandem compound steam engine which will soon be working again, powering 300 looms*

graze. It is the same with traction engines at a rally. You see these engines when they have knocked out a cog and they are just ticking over. People say, "Beautifully smooth." You connect the weight up and then listen to the row it makes, especially when they are knackered.

However, I have my own little stationary engine here in the garden and that really does do some work. It is not just a show piece. It may only have 90 odd foot of shafting to turn but at least you can tell that it is having to try to do it. You can put your hand on the shed and you get some idea of the effort that is involved to turn that all round. Whereas if you knock the belt off it runs as sweet as a nut.

## Queen Street Mill.

The use of steam in the weaving industry finally came to an end in March 1982 when Queen St. Mill at Harle Syke, near Burnley was closed. This was the last working steam powered textile mill in Britain and had remained virtually unchanged since it was built in 1894.

Its closure was no surprise in the hard hit textile trade but its claim to fame as the last steam powered textile mill ensured it received national news coverage. A major element of Britain's unrivalled industrial heritage appeared to be in danger of being lost for ever.

It is situated in the Harle Syke conservation area and is bordered by countryside and the unique qualities and character of the mill have attracted attention from both this country and overseas. Burnley Borough Council and Pennine Heritage Ltd have been working together to produce a scheme to save the buildings and contents with work now underway to open Queen St. Mill as a major tourist attraction. It was soon realised that it was essential for it to be retained as a *working* mill and not as a static museum piece.

Traditional products will be woven on the original Lancashire steam powered looms so the original 500HP steam engine will once again be in operation. It is hoped to have the mill open to visitors in Spring 1985 with a formal opening taking place in June.

Among the items I have come by over the years is a large ledger type book that has gold leaf letters on the front reading "Heron Mills Ltd, Engine Statistics". It is an amazing book for these days as it records every conceivable fact about the engines at the mill, entered every week by the engineman. Written in copperplate writing the statistics appear in columns. The first entry, which is typical of every subsequent week for many years reads:

1908
May 13th

| | | | |
|---|---|---|---|
| Coal Stock last week: | 35 tons | Revolutions: | 276,419 |
| Coal Received: | 62-15-2 | Hours + Mins: | 57.35 |
| Coal in Stock: | 35 tons | Cyl. Oil used Galls.: | 13 |
| Coal consumed: | 63 tons | Shafting Oil used Galls.: | 5 |
| I.H.P.* HP Cyl.: | 737 | Average temperature of | |
| I.H.P.* LP Cyl.: | 644 | water at economiser inlet: | 112 |
| I.H.P. Total: | 1381 | Average temperature of | |
| Lbs of Coal per I.H.P.: | 1.77 | water at economiser outlet: | 280 |
| | | General Remarks: | |

*Input Horse Power High Pressure
            Low Pressure

The General Remarks were entered as appropriate, usually referring to any faults, repairs and maintenance but some of these are:

17/6/08   Short time (The revolutions was down to 148,412 for the week).
30/10/12  New assistant started this morning.
7/11/12   New assistant got hit with crank. (Off work.)
21/11/12  Put fireman in Engine House and got a fresh fireman.
4/12/12   Very cold week and change of fireman.
30/4/13   Packers strike.
4/6/13    Packers strike settled May 31/13
16/7/13   Stopped Saturday for King & Queen visit to Oldham.
8/4/14    Yorkshire miners strike.
29/3/22   Mill stopped for engine repairs.

The last week of fully detailed entries was 29/5/28 then just half a dozen brief notes on some repairs in 1929 with nothing at all until 8/9/45. The columns of statistics were never resumed but notes were written in, across the printed columns describing any fault and repair, boiler washouts etc. This carries on like this until the last entry which is dated 26/6/62: "Got steam up at 5.55 engine started at 6.00. 6.30 pump broke but soon fixed. Engine still kept going. Stopped at 6.45".